PURE MATHEMATICS

7. INTEGRAL CALCULUS & APPLICATIONS

Second Edition

By

Anthony Nicolaides

P.A.S.S. PUBLICATIONS

Private Academic & Scientific Studies Limited

© A. NICOLAIDES 1994, 2007

First Published in Great Britain 1984 by
Private Academic & Scientific Studies Limited

ISBN–13   978–1–872684–73–4
SECOND EDITION 2008

This book is copyright under the Berne Convention.
All rights are reserved. Apart as permitted under the Copyright Act, 1956, no part of this publication may be reproduced, stored in a retrieval system, or transmitted in any form of by any means, electronic, electrical, mechanical, optical, photocopying, recording or otherwise, without the prior permission of the publisher.

Titles by the same author.
Revised and Enhanced

1. Algebra.   GCE A Level   ISBN–13   978–1–872684–82–6   £11–95

2. Trigonometry.   GCE A Level   ISBN–13   978–1–872684–87–1   £11–95

3. Complex Numbers.   GCE A Level   ISBN–13   978–1–872684–92–5   £9–95

4. Differential Calculus and Applications.   GCE A Level   ISBN–13   978–1–872684–97–0   £9–95

5. Cartesian and Polar Curve Sketching.   GCE A Level   ISBN–13   978–1–872684–63–5   £9–95

6. Coordinate Geometry in two Dimensions.   GCE A Level   ISBN–13   978–1–872684–68–0   £9–95

7. Integral Calculus and Applications.   GCE A Level   ISBN–13   978–1–872684–73–4   £14–95

8. Vectors in two and three dimensions.   GCE A Level   ISBN–13   978–1–872684–15–4   £9–95

9. Determinants and Matrices.   GCE A Level   ISBN–13   978–1–872684–16–1   £9–95

10. Probabilities.   GCE A Level   ISBN–13   978–1–872684–17–8   £8–95
    This book includes the full solutions

11. Success in Pure Mathematics: The complete works of GCE A Level. (1–9 above inclusive)   ISBN–13   978–1–872684–93–2   £39–95

12. Electrical & Electronic Principles. First year Degree Level   ISBN–13   978–1–872684–98–7   £16–95

13. GCSE Mathematics Higher Tier Third Edition.   ISBN–13   978–1–872684–69–7   £19–95

All the books have answers and a CD is attached with FULL SOLUTIONS of all the exercises set at the end of the book.

Preface

This book, which is part of the GCE A level series in Pure Mathematics covers the specialized topic of Integral Calculus and Applications.

The GCE A level series success in Pure Mathematics is comprised of nine books, covering the syllabuses of most examining boards. The books are designed to assist the student wishing to master the subject of Pure Mathematics. The series is easy to follow with minimum help. It can be easily adopted by a student who wishes to study it in the comforts of his home at his pace without having to attend classes formally; it is ideal for the working person who wishes to enhance his knowledge and qualification. Integral Calculus and Applications book, like all the books in the series, the theory is comprehensively dealt with, together with many worked examples and exercises. A step by step approach is adopted in all the worked examples. A CD is attached to the book with FULL SOLUTIONS of all the exercises set at the end of each chapter.

This book develops the basic concepts and skills that are essential for the GCE A level in Pure Mathematics.

C_1, C_2, C_3, C_4, FP1, FP2, FP3 are covered adequately in this book.

A. Nicolaides

7. INTEGRAL CALCULUS AND APPLICATIONS
CONTENTS
INTEGRAL CALCULUS

1. Integral Calculus

Introduction	1
Integral Sign	1
Algebraic Functions	2
SIMPLE DIFFERENTIAL EQUATIONS	3
Geometrical interpretation of the Arbitrary Constant of Integration	4
Displacement Velocity, Acceleration	6
Definite Integrals	8
Area under the Curve	9
Sign of the Area under the Curve	10
To Determine the Area under the Curve	12
Exercises 1	12

2. Trigonometric or Circular Functions — 15

Exercises 2 — 17

3. Exponential Functions — 19

Exercises 3 — 20

4. Logarithmic Functions — 21

The Integral of $\dfrac{1}{x}$ — 21

Exercises 4 — 22

5. Integration by Inspection — 25

Exercises 5 — 27

6. Hyperbolic Functions — 29

Definitions	29
Trigonometric and Hyperbolic Identities	30
Graph of Hyperbolic Functions	31
Inverse Hyperbolic Functions	32

Inverse Hyperbolic Functions	33
Hyperbolic Functions	34
Indefinite Integrals of Hyperbolic Functions	34
The Integration of the Squares of the Hyperbolic Function	36
The Integration of the Product of Hyperbolic Functions	37
Exercises 6	43

7. Integration by Parts — 46
Exercises 7 — 50

8. Reduction Formulae — 52
Definite Integrals — 57
Reduction Formulae — 60
Exercises 8 — 61

9. Approximate Numerical Integration — 63
Derivation of the Trapezoidal Rule — 63
Derivation of the Mid-ordinate Rule — 64
Simpson's Rule (Derivation) — 66
Exercises 9 — 69

10. Determines the Volume of Revolution of a Simple Area and locates the position of its Centroid — 71
Exercises 10 — 75

11. Differential Equations — 77
Separable Variables — 77
Exercises 11 — 77

12. Exact Differential Equations — 79
Exercises 12 — 79

13. The Integrating Factor — 80
Exercises 13 — 83

14. Second Order Differential Equations — 84
Equal and Different Roots — 85
Complex Roots — 85
Special Cases of the Second Order Differential Equations — 86
Exercises 14 — 90

— GCE A level

15. Length of Arc and Surface of Revolution — 92
Length of Arc in Cartesian Coordinates — 92
Length of Arc in Polar Coordinates — 92
Length of Arc Using Parametric Equations — 93
Surfaces — 98
Exercises — 102

16. Integration Using t-Formulae — 103
Hyperbolic Functions — 103
Circular Functions — 107
Exercises 16 — 109

17. The Mean or Average Values of Functions Over a Given Range — 111
Root Mean Square (R.M.S.) — 112
Exercises 17 — 115
Miscellaneous — 116
General Standard Integrals Answers — 128
Full Solutions CD — 1–98

Integral Calculus

Introduction

INTEGRAL CALCULUS, branch of infinitesimal calculus dealing with integrals of functions.

INTEGRATION is the process of finding the integral of a function. To integrate means to find the integral of, to combine parts into a whole, or to find the sum of the infinitesimally small strips, or elemental volumes.

Let us consider an example that illustrates integration.

Sketch the graph $y = x^2$ and find the area under the curve between $x = 1$ and $x = 2$, or the hatched area.

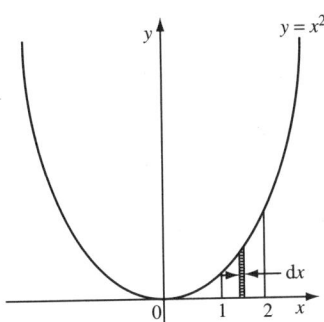

Fig. 7-I/1 Area under the graph.

Consider an elemental strip of width dx and ordinate y, the area of this elemental strip is $y dx$.

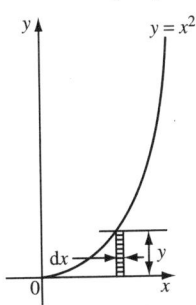

Fig. 7-I/2 An elemental strip of area $y dx$.

It is required to sum up all these strips that lie between $x = 1$ and $x = 2$, in other words to find the sum of the infinitesimally small strips between the limits $x = 1$ and $x = 2$.

Integral Sign

The integral sign $\int$ is evolved probably from the Greek letter Σ which denotes summation in Mathematics.

$\int_{x=1}^{x=2} y \, dx$ means to sum up all the elemental strips such as $y dx$ between the limits $x = 1$ and $x = 2$, that is, find the hatched area of Fig. 7-I/1.

$\int y \, dx$ is the indefinite integral of y with respect to x,

$\int_{1}^{2} y \, dx$ is the definite integral, in the former we add an arbitrary constant in the latter the arbitrary constant is cancelled. The formula for an algebraic function for integration is

$$\boxed{y = \int ax^n \, dx = \frac{ax^{n+1}}{n+1}}$$
$$+ \text{ an arbitrary constant where } n \neq -1 \quad \cdots (1)$$

where a and n are constants and x is the variable.

If $y = \dfrac{ax^{n+1}}{n+1} + $ constant, then the derivative is

$$\frac{dy}{dx} = \frac{a(n+1)x^n}{n+1} = ax^n,$$

therefore integration is the reverse of differentiation.

1

$\dfrac{dy}{dx} = ax^n$ the derivative, or the differential coefficient or the primitive,

$$dy = ax^n \, dx$$

integrating both sides, we have

$$\int dy = \int ax^n \, dx$$

$$y = \int ax^n \, dx = \dfrac{ax^{n+1}}{n+1} + c$$

where c is an arbitrary constant.

The function ax^n is always followed by dx in the integral, which means that we require to integrate ax^n with respect to x. Other integrals, involving different variables are shown for clarity

(i) $\int t^3 \, dt$,

(ii) $\int z^{\frac{1}{2}} \, dz$,

(iii) $\int y^{\frac{3}{2}} \, dy$,

(iv) $\int u \, du$,

(v) $\int v \, dv$

observe carefully the notation and do not forget the arbitrary constant.

Worked Example 1

Determine the indefinite integrals above by applying equation (1).

Solution 1

(i) $\int t^3 \, dt = \dfrac{t^{3+1}}{3+1} + c = \dfrac{t^4}{4} + c$

(ii) $\int z^{\frac{1}{2}} \, dz = \dfrac{z^{\frac{1}{2}+1}}{\frac{1}{2}+1} + c = \dfrac{2}{3} z^{\frac{3}{2}} + c$

(iii) $\int y^{\frac{3}{2}} \, dy = \dfrac{y^{\frac{3}{2}+1}}{\frac{3}{2}+1} + c = \dfrac{2}{5} y^{\frac{5}{2}} + c$

(iv) $\int u \, du = \dfrac{u^{1+1}}{1+1} + c = \dfrac{1}{2} u^2 + c$

(v) $\int v \, dv = \dfrac{v^{1+1}}{1+1} + c = \dfrac{1}{2} v^2 + c.$

Let us now evaluate the definite integral $\int_1^2 x^2 \, dx$ which denotes the area under the curve $y = x^2$ between the limits $x = 2$ and $x = 1$.

$$\int_1^2 x^2 \, dx = \left[\dfrac{x^3}{3} + c\right]_1^2$$

$$= \left[\dfrac{2^3}{3} + c\right] - \left[\dfrac{1^3}{3} + c\right]$$

$$= \left(\dfrac{8}{3} + c\right) - \left(\dfrac{1}{3} + c\right)$$

$$= \dfrac{7}{3} \text{ (square units)}$$

hence the arbitrary constant is cancelled.

As it can be seen, there was no need to include the arbitrary constant when evaluating a definite integral.

Algebraic Functions
Integration of the Form ax^n

Worked Example 2

$$\int x^{\frac{7}{3}} \, dx = \dfrac{x^{\frac{7}{3}+1}}{\frac{7}{3}+1} + c = \dfrac{3}{10} x^{\frac{10}{3}} + c.$$

Worked Example 3

$$\int \dfrac{1}{t^5} \, dt = \int t^{-5} \, dt$$

$$= \dfrac{t^{-5+1}}{-5+1} + c = -\dfrac{1}{4} t^{-4} + c$$

$$= -\dfrac{1}{4t^4} + c$$

the function $\dfrac{1}{t^5}$ is written in the form ax^n first and then we integrate.

Worked Example 4

$$\int (3x^2 + 5x - 1)\,dx = \int 3x^2\,dx + \int 5x\,dx - \int dx$$

$$= \frac{3x^3}{3} + c_1 + \frac{5x^2}{2}$$

$$+ c_2 - \frac{x^{0+1}}{0+1} - c_3$$

$$= x^3 + \frac{5}{2}x^2 - x + c$$

where $c = c_1 + c_2 - c_3$.

Worked Example 5

$$\int (v^3 + 5v^2 - 3v + 7)\,dv$$

$$= \int v^3\,dv + 5\int v^2\,dv - 3\int v\,dv + 7\int dv$$

$$= \frac{v^4}{4} + c_1 + \frac{5}{3}v^3 + c_2 - \frac{3}{2}v^2 - c_3 + 7v + c_4$$

$$= \frac{v^4}{4} + \frac{5}{3}v^3 - \frac{3}{2}v^2 + 7v + c$$

where $c = c_1 + c_2 - c_3 + c_4$.

Worked Example 6

$$\int \frac{y^5 + y^4}{2y}\,dy = \int \frac{y^5}{2y}\,dy + \int \frac{y^4}{2y}\,dy$$

$$= \frac{1}{2}\int y^4\,dy + \frac{1}{2}\int y^3\,dy$$

$$= \frac{1}{2}\frac{y^5}{5} + \frac{1}{2}\frac{y^4}{4} + c$$

$$= \frac{1}{10}y^5 + \frac{1}{8}y^4 + c.$$

In the above examples, we have integrated several terms and it is observed that each term integrated separately, there is of course an arbitrary constant for each individual integral, all these constants are lumped together to a single arbitrary constant.

Worked Example 7

$$\int \frac{x^2 + 3}{\sqrt{x}}\,dx = \int \left(\frac{x^2}{x^{\frac{1}{2}}} + \frac{3}{x^{\frac{1}{2}}}\right)dx$$

$$= \int \left(x^{\frac{3}{2}} + 3x^{-\frac{1}{2}}\right)dx$$

$$= \frac{2}{5}x^{\frac{5}{2}} + 6x^{\frac{1}{2}} + c.$$

Worked Example 8

$$\int \left(\frac{1}{t^3} + \frac{1}{t^5}\right)dt = \int t^{-3}\,dt + \int t^{-5}\,dt$$

$$= -\frac{1}{2}t^{-2} - \frac{1}{4}t^{-4} + c$$

$$= -\frac{1}{2t^2} - \frac{1}{4t^4} + c.$$

It is observed that the above integrals are evaluated by writing each term in the form of ax^n where n is positive or negative integer or rational.

Worked Example 9

Integrate $5\sqrt{x} + \sqrt[5]{x^3}$ with respect to x.

Solution 9

$$\int \left(5\sqrt{x} + \sqrt[5]{x^3}\right)dx = \int \left(5x^{\frac{1}{2}} + x^{\frac{3}{5}}\right)dx$$

$$= \frac{5x^{\frac{3}{2}}}{\frac{3}{2}} + \frac{x^{\frac{8}{5}}}{\frac{8}{5}} + c$$

$$= \frac{10}{3}x^{\frac{3}{2}} + \frac{5}{8}x^{\frac{8}{5}} + c.$$

Simple Differential Equations

Worked Example 10

If $\dfrac{dy}{dx} = -1$ and $y = 2$ when $x = 1$, find y in terms of x.

Solution 10

If $\dfrac{dy}{dx} = -1$, $dy = -dx$, integrating both sides

4 — GCE A level

$$\int dy = -\int dx$$

$$y = -x + c$$

so when $y = 2$, $x = 1$ then $2 = -1 + c$, $c = 3$
therefore

$$\boxed{y = -x + 3}$$

the arbitrary constant has been evaluated.

WORKED EXAMPLE 11

The gradient of a straight line is 5 and the line passes through the point $(3, 4)$, determine the equation of the line.

Solution 11

$$\frac{dy}{dx} = 5, \ dy = 5\, dx, \text{ integrating both sides}$$

$$\int dy = 5 \int dx$$

$$y = 5x + c$$

this line passes through the point $(3, 4)$, that is, $x = 3$ and $y = 4$ \quad $4 = 5(3) + c$

$c = 4 - 15 = -11$ and therefore

$$\boxed{y = 5x - 11}$$

WORKED EXAMPLE 12

The gradient of a curve at any point is given by $\frac{dy}{dx} = 3x^3 + 5x^2 - x + 2$. Find the equation of the curve, given that $y = 1$ when $x = 0$.

Solution 12

$$\frac{dy}{dx} = 3x^3 + 5x^2 - x + 2$$

$$dy = \left(3x^3 + 5x^2 - x + 2\right) dx$$

$$\int dy = \int \left(3x^3 + 5x^2 - x + 2\right) dx$$

$$y = \frac{3x^4}{4} + \frac{5x^3}{3} - \frac{x^2}{2} + 2x + c$$

General Solution
when $x = 0$, $y = 1$

$$1 = c$$

therefore the curve is given by

$$y = \frac{3}{4}x^4 + \frac{5}{3}x^3 - \frac{1}{2}x^2 + 2x + 1.$$

Geometrical Interpretation of the Arbitrary Constant of Integration

WORKED EXAMPLE 13

The gradient of a straight line is $\frac{dy}{dx} = 2$. Find the equations of the lines when the line passes through

(i) $(1, 0)$

(ii) $(0, 0)$

(iii) $(-3, 4)$. Draw these lines.

Solution 13

$$\frac{dy}{dx} = 2, \ dy = 2\, dx,$$

$$\int dy = \int 2\, dx, \ y = 2x + c,$$

the general equation of the line

(i) $x = 1, y = 0$

$$y = 2x + c$$

$$0 = 2(1) + c$$

$$c = -2$$

$$y = 2x - 2$$

the particular equation of the line

(ii) $x = 0, y = 0$

$$y = 2x + c$$

$$0 = 0 + c$$

general solution

$$c = 0$$

$$y = 2x$$

particular solution

(iii) $x = -3, y = 4$

$y = 2x + c$

general solution

$4 = 2(-3) + c$

$c = 10$

$y = 2x + 10$

the particular solution

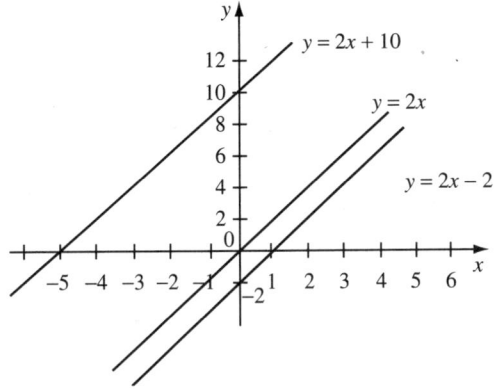

Fig. 7-I/3 The straight lines have the same gradient but intersect the x and y axes at different points.

We have noticed that in order to determine the constant further data is required.

WORKED EXAMPLE 14

The gradient of a curve is given by $\frac{dy}{dx} = 2x - 2$. Determine the equation of the curve when the curve cuts the y-axis at the points

(i) 4

(ii) 0

(iii) -2. Sketch the family of curves.

Solution 14

$\frac{dy}{dx} = 2x - 2, \quad dy = (2x - 2)\,dx,$

$\int dy = \int (2x - 2)\,dx$

$\boxed{y = x^2 - 2x + c}$

(i) when $x = 0, y = 4$ and hence $c = 4$
$y = x^2 - 2x + 4$

(ii) when $x = 0, y = 0$ and hence $c = 0$
$y = x^2 - 2x$

(iii) when $x = 0, y = -2$ and hence $c = -2$
$y = x^2 - 2x - 2$.

All the curves have a minimum since the coefficient of x^2 is positive, the minimum occurs at $x = -\frac{b}{2a}$

(i) $x = \frac{-(-2)}{2(1)} = 1, y_{\min} = 1 - 2 + 4 = 3$

(ii) $x = 1, y_{\min} = 1 - 2 = -1$

(iii) $x = 1, y_{\min} = 1 - 2 - 2 = -3$.

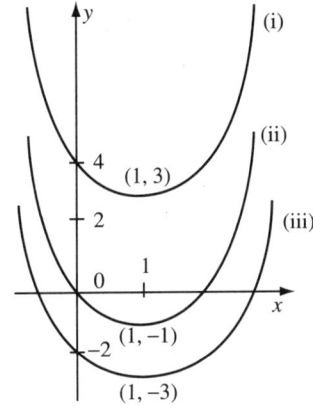

Fig. 7-I/4 The quadratic functions pass through different points.

The quadratic functions have similar shapes but the minima occur at different points and they intersect the x and y axes at different points.

WORKED EXAMPLE 15

Solve the following simple differential equations:

(i) $\frac{dy}{dx} = 5x$

(ii) $\frac{dy}{dx} = 3ax^2$

(iii) $\frac{dy}{dx} = 3x - 3$

(iv) $\frac{ds}{dt} = u + at$

(v) $\frac{dx}{dt} = \frac{1}{t^2} - \frac{1}{t^3} + \frac{1}{t^4}$.

Solution 15

(i) $\dfrac{dy}{dx} = 5x$

$$dy = 5x\,dx$$

$$\int dy = \int 5x\,dx$$

$$y = \dfrac{5}{2}x^2 + c$$

(ii) $\dfrac{dy}{dx} = 3ax^2$

$$dy = 3ax^2\,dx$$

$$\int dy = \int 3ax^2\,dx$$

$$y = ax^3 + c$$

(iii) $\dfrac{dy}{dx} = 3x - 3$

$$dy = (3x - 3)\,dx$$

$$\int dy = \int (3x - 3)\,dx$$

$$y = \dfrac{3}{2}x^2 - 3x + c$$

(iv) $\dfrac{ds}{dt} = u + at$

$$ds = (u + at)\,dt$$

$$\int ds = \int (u + at)\,dt$$

$$s = ut + \dfrac{at^2}{2} + c$$

(v) $\dfrac{dx}{dt} = \dfrac{1}{t^2} - \dfrac{1}{t^3} + \dfrac{1}{t^4}$

$$dx = \left(t^{-2} - t^{-3} + t^{-4}\right)dt$$

$$\int dx = \int \left(t^{-2} - t^{-3} + t^{-4}\right)dt$$

$$x = \dfrac{t^{-1}}{-1} - \dfrac{t^{-2}}{-2} + \dfrac{t^{-3}}{-3} + c$$

$$= -\dfrac{1}{t} + \dfrac{1}{2t^2} - \dfrac{1}{3t^3} + c.$$

Displacement Velocity Acceleration

$v = \text{velocity} = \dfrac{\text{displacement}}{\text{time}} = \dfrac{ds}{dt}$ (m/s)

$a = \text{acceleration} = \dfrac{\text{velocity}}{\text{time}} = \dfrac{dv}{dt}$ (m/s²)

$s = \text{displacement}$ (m).

WORKED EXAMPLE 16

The acceleration of a particle moving in a straight line is given by $a = 3t + 5$.

Determine

(i) the velocity

(ii) the displacement

given that $s = 0$ m and $v = 5$ m/s when $t = 0$ s.

Solution 16

(i) $a = 3t + 5$

$$a = \dfrac{dv}{dt} = 3t + 5$$

$$\int dv = \int (3t + 5)\,dt$$

$$v = \dfrac{3t^2}{2} + 5t + k$$

$v = 5$ when $t = 0$ $\quad 5 = k$

$$v = \dfrac{3}{2}t^2 + 5t + 5$$

$$\dfrac{ds}{dt} = \dfrac{3}{2}t^2 + 5t + 5$$

(ii) $\left|\int ds = \int \left(\dfrac{3}{2}t^2 + 5t + 5\right)dt\right|$

$$s = \dfrac{3}{2}\dfrac{t^3}{3} + \dfrac{5t^2}{2} + 5t + k$$

when $s = 0$, $t = 0$ then $k = 0$,

and $s = \dfrac{1}{2}t^3 + \dfrac{5}{2}t^2 + 5t.$

Substitution Method. Change of Variable

Determine $\int (ax+b)^n \, dx$.

Let $u = ax + b$, $\dfrac{du}{dx} = a$, $dx = \dfrac{du}{a}$

$$\int (ax+b)^n \, dx = \int u^n \dfrac{du}{a} = \dfrac{1}{a} \int u^n \, du$$

$$= \dfrac{u^{n+1}}{a(n+1)} + c$$

$$\boxed{\int (ax+b)^n \, dx = \dfrac{(ax+b)^n}{a(n+1)} + c.}$$

provided $n \ne -1$.

WORKED EXAMPLE 17

Find the indefinite integral $\int \dfrac{1}{(5x+7)^4} \, dx$.

Solution 17

$$\int \dfrac{1}{(5x+7)^4} \, dx = \int (5x+7)^{-4} \, dx$$

$$= \dfrac{(5x+7)^{-3}}{(-3) \times 5} + c$$

$$= -\dfrac{1}{15(5x+7)^3} + c$$

$\int \dfrac{1}{\sqrt{a^2 - x^2}} \, dx$

Let $x = a \sin \theta$ since $\sin^2 \theta + \cos^2 \theta = 1$

$\cos^2 \theta = 1 - \sin^2 \theta$

$\cos \theta = \sqrt{1 - \sin^2 \theta}$

$\dfrac{dx}{d\theta} = a \cos \theta$

$$\int \dfrac{1}{\sqrt{a^2 - a^2 \sin^2 \theta}} a \cos \theta \, d\theta$$

$$= \int d\theta = \theta = \sin^{-1} \dfrac{x}{a} + c$$

$$\boxed{\int \dfrac{1}{\sqrt{a^2 - x^2}} \, dx = \sin^{-1} \dfrac{x}{a} + c}$$

Alternatively $\int \dfrac{1}{\sqrt{a^2 - x^2}} \, dx$

Let $x = a \cos \theta$, $\dfrac{dx}{d\theta} = -a \sin \theta$,

$dx = -a \sin \theta \, d\theta$

$$\int \dfrac{1}{\sqrt{a^2 - x^2}} \, dx = \int \dfrac{-a \sin \theta \, d\theta}{\sqrt{a^2 - \cos^2 \theta}}$$

$$= -\int d\theta = -\theta + c$$

$$\int \dfrac{1}{\sqrt{a^2 - x^2}} \, dx$$

$$= -\cos^{-1} \dfrac{x}{a} + c.$$

or $\boxed{\int \dfrac{dx}{\sqrt{a^2 - x^2}} = -\cos^{-1} \dfrac{x}{a} + c}$

$\int \dfrac{1}{\sqrt{a^2 + x^2}} \, dx$

Let $x = a \tan \theta$, $\dfrac{dx}{d\theta} = a \sec^2 \theta$,

$dx = a \sec^2 \theta \, d\theta$

$$\int \dfrac{1}{a^2 + x^2} \, dx = \int \dfrac{1}{a^2 + a^2 \tan^2 \theta} a \sec^2 \theta \, d\theta$$

$$= \dfrac{1}{a} \int d\theta = \dfrac{1}{a} \theta + c$$

since $1 + \tan^2 \theta = \sec^2 \theta$

$$\boxed{\int \dfrac{1}{a^2 + x^2} \, dx = \dfrac{1}{a} \tan^{-1} \dfrac{x}{a} + c}$$

WORKED EXAMPLE 18

Find the definite integrals:

(i) $\int \dfrac{1}{\sqrt{1 - x^2}} \, dx$

(ii) $\int \dfrac{1}{1 + x^2} \, dx$

(iii) $\int \dfrac{1}{\sqrt{1 - 4x^2}} \, dx$

(iv) $\int \dfrac{1}{4+x^2}\,dx$

(v) $\int \dfrac{1}{9+x^2}\,dx$

using the standard integrals derived.

Solution 18

(i) $\int \dfrac{1}{\sqrt{1-x^2}}\,dx = \sin^{-1}\dfrac{x}{1} + c$ since $a=1$

(ii) $\int \dfrac{1}{1+x^2}\,dx = \tan^{-1} x + c$ where $a=1$

(iii) $\int \dfrac{1}{\sqrt{1-4x^2}}\,dx = \int \dfrac{1}{2\sqrt{\left(\frac{1}{2}\right)^2 - x^2}}\,dx$

$= \dfrac{1}{2}\sin^{-1}\dfrac{x}{\frac{1}{2}} + c = \dfrac{1}{2}\sin^{-1} 2x + c$

(iv) $\int \dfrac{1}{4+x^2}\,dx = \int \dfrac{1}{2^2 + x^2}\,dx = \dfrac{1}{2}\tan^{-1}\dfrac{x}{2} + c$

(v) $\int \dfrac{1}{9+x^2}\,dx = \int \dfrac{1}{3^2 + x^2}\,dx = \dfrac{1}{3}\tan^{-1}\dfrac{x}{3} + c.$

Definite Integrals

A definite integral is an integral evaluated between two limits and has a finite answer, the arbitrary constant of integration is cancelled. There are two limits the upper limit and the lower limit $\left(\int_{lower}^{upper}\right)$.

WORKED EXAMPLE 19

Evaluate the definite integrals:

(i) $\int_2^8 2x\,dx$

(ii) $\int_1^3 3x^2\,dx$

(iii) $\int_0^1 \left(5x^2 + x + 5\right) dx$

(iv) $\int_{1.5}^{3.5} (3x-1)^3\,dx$

(v) $\int_4^5 (x+1)(x-3)\,dx$

(vi) $\int_0^3 3x^3\,dx$

(vii) $\int_2^4 4x^5\,dx$

(viii) $\int_0^1 \dfrac{5x^2 - x}{x}\,dx$

(ix) $\int_0^1 \left(\dfrac{x^5}{5} + \dfrac{x^4}{4} + \dfrac{x^3}{3}\right) dx$

(x) $\int_1^2 \left(\sqrt{x} + x^{-\frac{1}{2}}\right) dx.$

Solution 19

(i) $\int_2^8 2x\,dx = \left[\dfrac{2x^2}{2}\right]_2^8 = \left[x^2\right]_2^8 = \left[8^2\right] - \left[2^2\right]$

$= 64 - 4 = 60$ square units

(ii) $\int_1^3 3x^2\,dx = \left[\dfrac{3x^3}{3}\right]_1^3 = \left[3^3\right] - \left[1^3\right]$

$= 27 - 1 = 26$ square units

(iii) $\int_0^1 \left(5x^2 + x + 5\right) dx = \left[\dfrac{5x^3}{3} + \dfrac{x^2}{2} + 5x\right]_0^1$

$= \left[\dfrac{5}{3}(1)^3 + \dfrac{1^2}{2} + 5(1)\right]$

$- \left[\dfrac{5}{3}(0)^3 + \dfrac{1}{2}(0)^2 + 5(0)\right]$

$= \dfrac{5}{3} + \dfrac{1}{2} + 5 = 7\dfrac{1}{6}$ square units

(iv) $\int_{1.5}^{3.5}(3x-1)^3 \, dx = \left[\frac{(3x-1)^4}{4\times 3}\right]_{1.5}^{3.5}$

$= \left[\frac{(3(3.5)-1)^4}{12}\right]$

$- \left[\frac{(3(1.5)-1)^4}{12}\right]$

$= \frac{9.5^4 - 3.5^4}{12}$

$= 666$ square units to 3 s.f.

(v) $\int_{4}^{5}(x+1)(x-3) \, dx$

$= \int_{4}^{5}\left(x^2 + x - 3x - 3\right) dx$

$= \int_{4}^{5}\left(x^2 - 2x - 3\right) dx = \left[\frac{x^3}{3} - x^2 - 3x\right]_{4}^{5}$

$= \left[\frac{5^3}{3} - 5^2 - 3(5)\right] - \left[\frac{4^3}{3} - 4^2 - 3(4)\right]$

$= 1.67 - (-6.667)$

$= 8.34$ square units to 3 s.f.

(vi) $\int_{0}^{3} 3x^3 \, dx = \left[\frac{3}{4}x^4\right]_{0}^{3} = \left[\frac{3}{4}(3)^4\right] - \left[\frac{3}{4}(0)^4\right]$

$= 60.75$ square units

(vii) $\int_{2}^{4} 4x^5 \, dx = \left[\frac{4}{6}x^6\right]_{2}^{4} = \left[\frac{2}{3}(4)^6\right] - \left[\frac{2}{3}(2)^6\right]$

$= 2730.7 - 42.7 = 2688$ square units

(viii) $\int_{0}^{1} \frac{5x^2 - x}{x} \, dx = \int_{0}^{1} (5x-1) \, dx$

$= \left[\frac{5x^2}{2} - x\right]_{0}^{1} = \frac{5}{2} - 1 = \frac{3}{2}$

(ix) $\int_{0}^{1}\left(\frac{x^5}{5} - \frac{x^4}{4} + \frac{x^3}{3}\right) dx = \left[\frac{x^6}{30} - \frac{x^5}{20} + \frac{x^4}{12}\right]_{0}^{1}$

$= \frac{1}{30} - \frac{1}{20} + \frac{1}{12} = \frac{1}{15}$ square units

(x) $\int_{1}^{2}\left(\sqrt{x} + x^{-\frac{1}{2}}\right) dx$

$= \int_{1}^{2}\left(x^{\frac{1}{2}} + x^{-\frac{1}{2}}\right) dx = \left[\frac{x^{\frac{3}{2}}}{\frac{3}{2}} + \frac{x^{\frac{1}{2}}}{\frac{1}{2}}\right]_{1}^{2}$

$= \left[\frac{2}{3}(2)^{\frac{3}{2}} + 2(2)^{\frac{1}{2}}\right] - \left[\frac{2}{3} + 2\right]$

$= 1.886 + 2.828 - 2.667$

$= 2.047$ square units to 3 d.p.

Area under the Curve

We have already seen a simple example at the beginning at this chapter introducing integration.

The area under the curve of the simple algebraic function $y = x^2$ between the limit $x = 1$ and $x = 2$.

$\int_{1}^{2} x^2 \, dx = \left[\frac{x^3}{3}\right]_{1}^{2} = \left[\frac{2^3}{3}\right] - \left[\frac{1^3}{3}\right] = \frac{8}{3} - \frac{1}{3}$

$= \frac{7}{3}$ square units

this of course is called a definite integral.

WORKED EXAMPLE 20

Determine the area under the curve $y = x^2 + 1$ between the coordinate axes and the line $x = 2$.

Solution 20

It is obviously required to sketch this curve so that we can work out clearly the limits of integration.

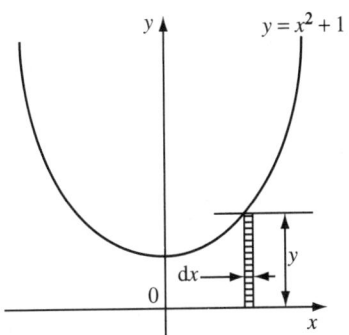

Fig. 7-I/5 The area under the curve.

Consider an elemental strip of width dx and an ordinate y, summing up all these strips between $x = 0$ and $x = 2$.

$$\int_0^2 (x^2 + 1)\, dx = \left[\frac{x^3}{3} + x\right]_0^2$$

$$= \frac{2^3}{3} + 2 - 0$$

$$= \frac{8}{3} + 2 = 4\frac{2}{3}.$$

WORKED EXAMPLE 21

Determine the area under the curve $y = x^2 - 5$ between the coordinate axes and the line $x = 2$.

Solution 21

The graph of $y = x^2 - 5$ is shown in Fig. 7-I/6.

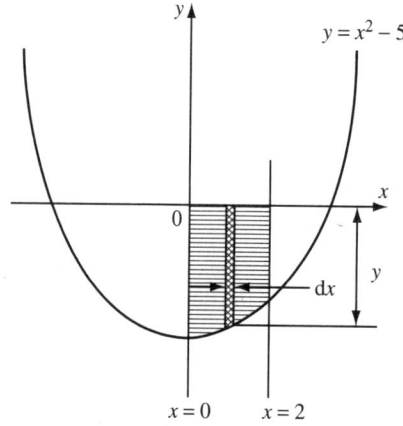

Fig. 7-I/6 The area under the curve

The elemental area $= y\, dx$

$$\int_0^2 (x^2 - 5)\, dx = \left[\frac{x^3}{3} - 5x\right]_0^2$$

$$= \frac{8}{3} - 10$$

$$= -\frac{22}{3}.$$

The area evaluated is negative, it is observed that the area below the x-axis will have a negative sign and the area above the x-axis will have a positive sign.

Sign of the Area under the Curve

The area above the x-axis is positive and the area below the x-axis is negative.

WORKED EXAMPLE 22

Sketch the graphs:

(i) $y^2 = 4x$

(ii) $y^2 = -4x$

(iii) $x^2 = 4y$

(iv) $x^2 = -4y$.

Determine the area enclosed by the four curves.

Solution 22

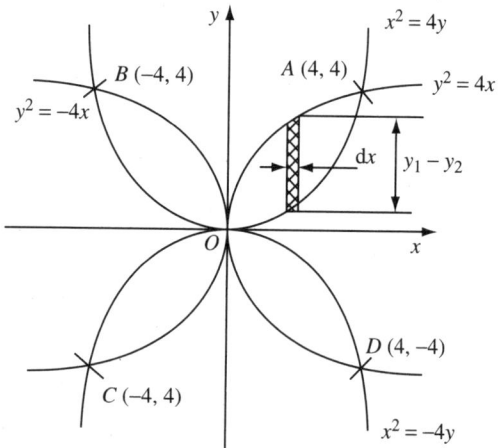

Fig. 7-I/7 The graphs

Fig. 7-I/7 shows the sketch of these graphs which are easily obtained.

$y^2 = 4x$, if $x = 0, y = 0$; if $x = 1$, $y = \pm 2$

$x^2 = 4y$, if $y = 0, x = 0$; if $y = 1$, $x = \pm 2$

$y^2 = -4x$, if $x = 0, y = 0$; if $x = -1$, $y = \pm 2$

$x^2 = -4y$, if $y = 0, x = 0$; if $y = -1$, $x = \pm 2$.

Solving the equations $x^2 = 4y$ and $y^2 = 4x$ simultaneously, we find the intersection at O and A.

$$y^2 = 4x = \left(\frac{x^2}{4}\right)^2 = \frac{x^4}{16}$$

$$x^4 = 64x \quad x(x^3 - 64) = 0$$

$$x = 0 \text{ or } x = 4 \quad \text{and} \quad y = 0 \text{ or } y = 4.$$

Similarly, we find the intersections at B, C, and D respectively.

There are four loops of areas to evaluate. Consider the loop enclosed by $y^2 = 4x$ and $x^2 = 4y$, consider a strip of width dx and height $y_1 - y_2$ where $y_1^2 = 4x$ and $4y_2 = x^2$. The area of the loop enclosed between $y^2 = 4x$ and $x^2 = 4y$.

$$= \int_0^4 (y_1 - y_2)\, dx = \int_0^4 \left(\sqrt{4x} - \frac{x^2}{4}\right) dx$$

$$= \int_0^4 \left(2x^{\frac{1}{2}} - \frac{1}{4}x^2\right) dx = \left[\frac{2x^{\frac{3}{2}}}{\frac{3}{2}} - \frac{1}{12}x^3\right]_0^4$$

$$= \frac{4}{3}4\sqrt{4} - \frac{64}{12} = \frac{32}{3} - 5.33$$

$$= 10.67 - 5.33 = 5.34 \text{ square units.}$$

Similarly the area between $y^2 = -4x$ and $x^2 = 4y$ is 5.34 square units. The areas between $x^2 = 4y$ and $y^2 = 4x$; and $x^2 = -4y$ and $y^2 = -4x$ are given respectively -5.34 and -5.34 square units. The areas above the x-axis are positive and the areas below the x-axis are negative, therefore the total area of the four loops will be four times that of 5.34 square units, the significance of the negative area is that it lies below the x-axis.

Total area = $5.34 \times 4 = 21.36$ square units.

WORKED EXAMPLE 23

Determine the area between $x^2 = y$ and the line $y = 2x$.

Solution 23

Sketch the line $y = 2x$ and the curve $x^2 = y$, as shown in Fig. 7-I/8.

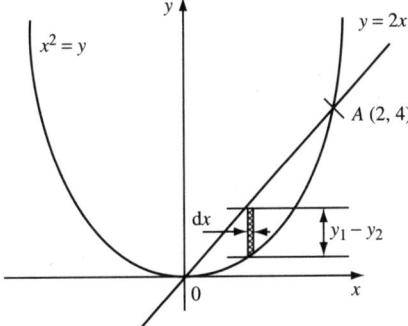

Fig. 7-I/8

$x^2 = y$ and $y = 2x$ are solved simultaneously $x^2 = 2x$ $x(x - 2) = 0, x = 0$ or $x = 2$ then $y = 0$, or $y = 4$, therefore the intersections between the line and the curve are $(0, 0)$ and $(2, 4)$.

The area under the line between $x = 0$ and $x = 2$,

$$\int_0^2 2x\, dx = \left[\frac{2x^2}{2}\right]_0^2 = 4 \text{ square units alternatively the}$$

area of a right angled triangle $= \frac{1}{2} \times 2 \times 4 = 4$ square units.

The area under the curve $x^2 = y$

$$= \int_0^2 x^2\, dx = \left[\frac{x^3}{3}\right]_0^2 = \frac{8}{3}.$$

The area enclosed between the line and the curve

$$= \text{area of triangle} - \text{area under the curve } (x^2 = y)$$

$$= 4 - \frac{8}{3} = \frac{4}{3} \text{ square units.}$$

Consider the strip of height $y_1 - y_2$ and width dx

$$\int_0^2 (y_1 - y_2)dx = \int_0^2 (2x - x^2)dx = \left[\frac{2x^2}{2} - \frac{x^3}{3}\right]_0^2$$

$$= 4 - \frac{8}{3} = \frac{4}{3} \text{ square units}$$

To Determine the Area under the Curve

Calculate the area enclosed between the curve $y^2 = x$, the x-axis and $x = 3$.

Consider an elemental strip dx and height y, its area is $y\,dx$

$$\text{Area required} = \int_0^3 y\,dx = \int_0^3 \sqrt{x}\,dx$$

$$= \int_0^3 x^{\frac{1}{2}}\,dx = \left[\frac{x^{\frac{3}{2}}}{\frac{3}{2}}\right]_0^3$$

$$= \frac{2}{3}3\sqrt{3} = \frac{6}{3}\sqrt{3} = 2\sqrt{3} \text{ s.u.}$$

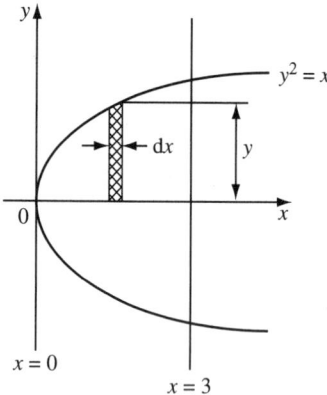

Fig. 7-I/9

Calculate now the area enclosed between the curve $y^2 = x$, the y-axis and $y = \sqrt{3}$.

Consider now the elemental strip with dy width and x height, its area is $x\,dy$.

Area required

$$= \int_0^{\sqrt{3}} x\,dy = \int_0^{\sqrt{3}} y^2\,dy = \left[\frac{y^3}{3}\right]_0^{\sqrt{3}} = \frac{3\sqrt{3}}{3} = \sqrt{3} \text{ s.u.}$$

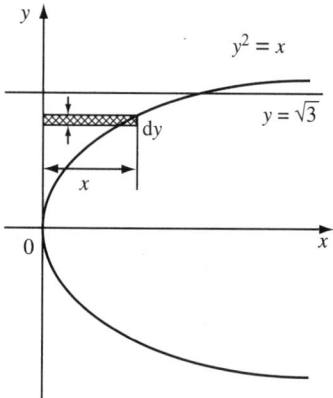

Fig. 7-I/10

Calculate the area enclosed between the curve $y^2 = x$, the y-axis and $y = x^2$.

Consider the dark strip whose area is $(y_1 - y_2)\,dx$.

Area required

$$= \int_0^1 \left(\sqrt{x} - x^2\right) dx = \int_0^1 \left(x^{\frac{1}{2}} - x^2\right) dx$$

$$= \left[\frac{x^{\frac{3}{2}}}{\frac{3}{2}} - \frac{x^3}{3}\right]_0^1 = \frac{2}{3} - \frac{1}{3} = \frac{1}{3} \text{ s.u.}$$

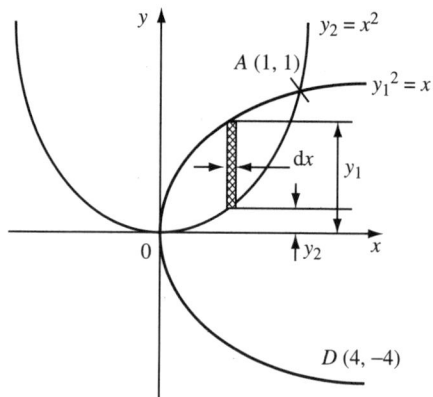

Fig. 7-I/11

Exercises 1

1. Integrate with respect to x the following:

   (i) x^3

   (ii) $3x^3 + 5x^2 - 7x$

   (iii) $\dfrac{1}{x^2}$

   (iv) $(x+1)(x+2)(x+3)$

   (v) $\dfrac{1}{x^{\frac{1}{2}}} + \dfrac{1}{x^{\frac{1}{3}}} + \dfrac{1}{x^{\frac{1}{4}}}$

   (vi) $\dfrac{x^2 + 3}{x^2}$

   (vii) $x^{\frac{3}{4}}$

   (viii) $\dfrac{x^3 + 2x^2 + 5x^{\frac{1}{2}} - 1}{x^2}$

   (ix) $\sqrt[5]{x^4}$

   (x) x^{-5}

   (xi) $\sqrt{1+x}$

   (xii) $\sqrt[3]{(x-1)^2}$

   (xiii) $\dfrac{1}{\sqrt{x}}$

   (xiv) $\dfrac{x-1}{\sqrt{x}}$

   (xv) $\dfrac{x^2 - x - 1}{\sqrt[3]{x^2}}$.

2. Integrate with respect to Z the following:

(i) $Z + \dfrac{1}{Z^3} + \dfrac{1}{Z^2}$

(ii) $\dfrac{Z}{1} + \dfrac{Z^2}{2} + \dfrac{Z^3}{3}$

(iii) $Z^n + Z^{-n}$

(iv) $Z^{\frac{3}{5}}$

(v) $aZ^2 + bZ + c$

(vi) $\dfrac{1}{Z^2} + \dfrac{1}{2Z^3} + \dfrac{1}{3Z^4}$

(vii) $Z^1 + Z^2 + Z^3 + Z^4$

(viii) $\dfrac{Z + Z^3}{Z^3}$

(ix) $(1 + Z)^3$

(x) $(1 + 2Z)^2$.

3. Integrate with respect to t the following:

(i) $1 + t$

(ii) 5

(iii) $1 + t^3$

(iv) $(1 + t)^2$

(v) $\dfrac{t^3 + t^2 + t^1}{3t}$

(vi) $3t + 5$

(vii) $\dfrac{1}{\sqrt[5]{t^2}}$

(viii) $\dfrac{1}{t^{\frac{2}{3}}}$

(ix) $\dfrac{1}{\sqrt[3]{t^{-2}}}$

(x) $\dfrac{t^2 + 1}{\sqrt{t}}$.

4. (i) $\displaystyle\int \left(3x^2 + \dfrac{1}{2}x - \dfrac{3}{x^3}\right) dx$

(ii) $\displaystyle\int \left(\dfrac{1}{\sqrt{x}} - \dfrac{1}{\sqrt[3]{x^2}}\right) dx$

(iii) $\displaystyle\int \left(x^4 + 5x^3 - 3x^2 + 7\right) dx$

(iv) $\displaystyle\int \left(t^{\frac{1}{2}} - t^{\frac{1}{3}} + t^{\frac{1}{4}}\right) dt$

(v) $\displaystyle\int \left(Z^{0.5} - Z^{0.75} + Z^{1.3}\right) dZ$

(vi) $\displaystyle\int \dfrac{y^2 + 3y - 5}{y^4} dy$

(vii) $\displaystyle\int \dfrac{(x^3 - 1)}{x^2} dx$

(viii) $\displaystyle\int \dfrac{t^4 - 3t^3 - 5}{t^2} dt$.

5. A line passes through two points $(3, 1)$ and $(-2, 4)$, write down the gradient in the form of a differential coefficient and hence determine the equation.

6. The gradient of a function is $\dfrac{dy}{dx} = x^3 - \dfrac{1}{x^2} + x$. The curve passes through the point $(1, 2)$, find the equation of the curve.

7. Solve the following simple differential equations:

(i) $\dfrac{dy}{dx} = \dfrac{5}{x^2}$

(ii) $\dfrac{dy}{dx} = \dfrac{1}{\sqrt{x}} - \sqrt{x} + x^{\frac{3}{2}}$

(iii) $\dfrac{dy}{dx} = ax^2 + bx + c$.

8. A particle moves along a straight line OA with velocity $(3t + 5)$ m/s.

Find an expression for its displacement from O in terms of t (the particle is at O when $t = 0$). Determine the displacement at $t = 1$ and $t = 5$s.

9. A stone is thrown vertically downwards at 100 m/s from the top of a cliff.
If the acceleration due to gravity is given as $\dfrac{dv}{dt} = 10$ m/s^2. Determine the velocity and displacement after t s.

10. Find the following definite integrals:

    (i) $\int \dfrac{1}{(5x+3)^2}\, dx$

    (ii) $\int (x-7)^{25}\, dx$.

11. (i) $\int \dfrac{1}{\sqrt{2-x^2}}\, dx$

    (ii) $\int \dfrac{1}{\sqrt{3-x^2}}\, dx$

    (iii) $\int \dfrac{1}{5^2+x^2}\, dx$

    (iv) $\int \dfrac{1}{4+x^2}\, dx$.

12. Calculate the area enclosed by the curve $y = x^2 - 4x + 5$, the x-axis and the lines $x = 3$ and $x = 4$.

13. Sketch the curves:

    (i) $y = 3x^2$ and (ii) $y = 5 - x^2$.

    Show that the coordinates of the points of intersection of (i) and (ii) are:

    $$A\left(-\dfrac{\sqrt{5}}{2},\, \dfrac{15}{4}\right) \quad \text{and} \quad B\left(\dfrac{\sqrt{5}}{2},\, \dfrac{15}{4}\right).$$

    Determine exactly the following:

    (a) The area between the curve (i), the x-axis and the lines
    $$x = -\dfrac{\sqrt{5}}{2} \quad \text{and} \quad x = \dfrac{\sqrt{5}}{2}.$$

    (b) The area between the curve (ii) and the x-axis.

    (c) The area between the curve (ii), the x-axis and the lines
    $$x = -\dfrac{\sqrt{5}}{2} \quad \text{and} \quad x = \dfrac{\sqrt{5}}{2}.$$

14. Determine the area enclosed by the parabolas $y = x^2 - 5x + 7$ and $y = -x^2 + 3x + 1$.

2

Trigonometric or Circular Functions

$$\int \cos x \, dx = \sin x + k$$

the derived function of $\sin x + k$ is $\cos x$.

$$\int \sin x \, dx = -\cos x + k$$

the derived function of $-\cos x + k$ is $\sin x$.

$$\int \sin ax \, dx = -\frac{\cos ax}{a} + k$$

the derived function of $-\frac{\cos ax}{a} + k$ is $\sin ax$.

$$\int \cos ax \, dx = \frac{\sin ax}{a} + k$$

the derived function of $\frac{\sin ax}{a} + k$ is $\cos ax$.

To show that $\int \sin ax \, dx = -\frac{\cos ax}{a} + k$.

Let $u = ax$, $\frac{du}{dx} = a$ hence $dx = \frac{du}{a}$

$$\int \sin u \frac{du}{a} = \frac{1}{a} \int \sin u \, du$$

$$= \frac{1}{a}(-\cos u) + k = -\frac{\cos ax}{a} + k$$

therefore

$$\boxed{\int \sin ax \, dx = -\frac{\cos ax}{a} + k}$$

Similarly to show that $\int \cos ax \, dx = \frac{\sin ax}{a} + k$

Let $u = ax$, $\frac{du}{dx} = a$, $dx = \frac{du}{a}$

$$\int \cos ax \, dx = \int \cos u \frac{du}{a}$$

$$= \frac{1}{a} \int \cos u \, du = \frac{1}{a}[\sin u] + k$$

$$\boxed{\int \cos ax \, dx = \frac{\sin ax}{a} + k}$$

WORKED EXAMPLE 24

$$\int \sin 3x \, dx = -\frac{\cos 3x}{3} + k.$$

WORKED EXAMPLE 25

$$\int \sin \frac{1}{2}x \, dx = -2 \cos \frac{1}{2}x + k.$$

WORKED EXAMPLE 26

$$\int 3 \sin 5x \, dx = -\frac{3 \cos 5x}{5} + k.$$

WORKED EXAMPLE 27

$$\int 5 \cos 5x \, dx = \sin 5x + k.$$

WORKED EXAMPLE 28

$$\int (3 \sin 2x - 4 \cos 3x) dx$$

$$= -\frac{3}{2} \cos 2x - \frac{4}{3} \sin 3x + k.$$

15

WORKED EXAMPLE 29

$$\int 7\cos(-5x)dx = \frac{7\sin(-5x)}{-5} + k$$

$$= \frac{7}{5}\sin 5x + k.$$

WORKED EXAMPLE 30

$$3\int \sin\left(-\frac{3}{4}x\right) dx = \frac{3\cos\left(-\frac{3x}{4}\right)}{-\frac{3}{4}} + k$$

$$= -4\cos\frac{3x}{4} + k.$$

WORKED EXAMPLE 31

$$\text{Evaluate} \int_0^{\frac{\pi}{2}} \cos 3x \, dx = \left[\frac{\sin 3x}{3}\right]_0^{\frac{\pi}{2}}$$

$$= \left[\frac{\sin \frac{3\pi}{2}}{3}\right] - \left[\frac{\sin 0}{3}\right] = -\frac{1}{3}.$$

WORKED EXAMPLE 32

$$\text{Evaluate} \int_0^{\frac{\pi}{4}} \sin 2x \, dx = \left[\frac{\cos 2x}{2}\right]_0^{\frac{\pi}{4}}$$

$$= \left[\frac{-\cos \frac{2\pi}{4}}{2}\right] - \left[\frac{-\cos 0}{2}\right]$$

$$= 0 + \frac{\cos 0}{2} = \frac{1}{2}.$$

WORKED EXAMPLE 33

$$\text{Evaluate} \int_{\frac{\pi}{6}}^{\frac{\pi}{3}} (\sin 2x + \cos 3x) dx$$

$$= \left[-\frac{\cos 2x}{2} + \frac{\sin 3x}{3}\right]_{\frac{\pi}{6}}^{\frac{\pi}{3}}$$

$$= \left[-\frac{\cos 2\left(\frac{\pi}{3}\right)}{2} + \frac{\sin 3\left(\frac{\pi}{3}\right)}{3}\right]$$

$$- \left[-\frac{\cos 2\left(\frac{\pi}{6}\right)}{2} + \frac{\sin 3\left(\frac{\pi}{6}\right)}{3}\right]$$

$$= -\frac{1}{2}\left(-\frac{1}{2}\right) + \frac{0}{3} + \frac{1}{4} - \frac{1}{3}$$

$$= \frac{1}{4} + \frac{1}{4} - \frac{1}{3}$$

$$= \frac{1}{2} - \frac{1}{3}$$

$$= \frac{3-2}{6}$$

$$= \frac{1}{6}.$$

Find the integrals

(i) $\int \sin^2 x \, dx$

(ii) $\int \cos^2 x \, dx$

(iii) $\int \sin^2 2x \, dx$

(iv) $\int \cos^2 2x \, dx$

(v) $\int \sin^2 \frac{1}{2}x \, dx.$

Solutions

(i) $\int \sin^2 x \, dx$ use the double angle formulae

$$\cos 2x = 2\cos^2 x - 1 = 1 - 2\sin^2 x$$

$$2\sin^2 x = 1 - \cos 2x$$

$$\sin^2 x = \frac{1}{2}(1 - \cos 2x)$$

$$\int \sin^2 x \, dx = \frac{1}{2} \int (1 - \cos 2x) dx$$

$$= \frac{1}{2}\left(x - \frac{\sin 2x}{2}\right) + c$$

(ii) $\int \cos^2 x \, dx$

$2\cos^2 x = \cos 2x + 1$

$\cos^2 x = \frac{1}{2}(\cos 2x + 1)$

$\int \cos^2 x \, dx = \frac{1}{2}\int (\cos 2x + 1) dx$

$= \frac{1}{2}\left(\frac{\sin 2x}{2} + x\right) + c$

(iii) $\int \sin^2 2x \, dx$

$\cos 4x = 1 - 2\sin^2 2x$

$\sin^2 2x = \frac{1 - \cos 4x}{2}$

$\int \sin^2 2x \, dx = \frac{1}{2}\int (1 - \cos 4x) dx$

$= \frac{1}{2}\left(x - \frac{\sin 4x}{4}\right) + c$

(iv) $\int \cos^2 2x \, dx = \frac{1}{2}\int (\cos 4x + 1) dx$

$= \frac{1}{2}\left(\frac{\sin 4x}{4} + x\right) + c$

$\cos 4x = 2\cos^2 2x - 1$

$\cos^2 2x = \frac{\cos 4x + 1}{2}$

(v) $\int \sin^2 \frac{1}{2}x \, dx = \frac{1}{2}\int (1 - \cos x) dx$

$= \frac{1}{2}(x - \sin x) + c$

$\cos x = 2\cos^2 \frac{x}{2} - 1$

$= 1 - 2\sin^2 \frac{x}{2}$

$\sin^2 \frac{x}{2} = \frac{1}{2}(1 - \cos x).$

Now think of the variable being different from x.

$\int \sin x \, d(\sin x) = \frac{\sin^2 x}{2} + c$

$\int \tan^2 x \, d(\tan x) = \frac{\tan^3 x}{3} + c$

$\int \cos(1 + x) \, d(1 + x) = \sin(1 + x) + c$

$\int \nabla \, d\nabla = \frac{\nabla^2}{2} + c$

$\int \varepsilon^{\frac{1}{2}} \, d\varepsilon = \frac{\varepsilon^{\frac{3}{2}}}{\frac{3}{2}} + c$

$\int \sqrt{\sin x} \, d(\sin x) = \frac{(\sin x)^{\frac{3}{2}}}{\frac{3}{2}} + c$

$\int \sin(5x - 7) \, d(5x - 7) = -\cos(5x - 7) + c$

$\int \cos(1 - 4x) \, d(1 - 4x) = \sin(1 - 4x) + c$

$\int \sin\left(\frac{1}{x}\right) d\left(\frac{1}{x}\right) = -\cos\left(\frac{1}{x}\right) + c$

all the above integrals use the formula

$\int ax^n dx = \frac{ax^{n+1}}{n+1} + c$

the variable are different from x.

Exercises 2

1. Integrate the following trigonometric functions:

   (i) $\sin kx$

   (ii) $\cos nx$

   (iii) $3 \sin \frac{1}{2}x$

   (iv) $-5 \sin 3x$

   (v) $-5 \cos 5x.$

2. Evaluate the integrals

(i) $\int_0^{\pi/2} \sin t \, dt$

(ii) $\int_{\pi/8}^{\pi/4} \cos t \, dt$

(iii) $\int_{\pi/3}^{\pi/2} \sin 2t \, dt.$

3. Evaluate the integrals

(i) $\int_0^{\pi} (3\sin y + 5\cos y) \, dy$

(ii) $\int_0^{2\pi} (5\cos 3y - 3\sin 2y) \, dy.$

4. Evaluate

(i) $\int (5\sin 3x - \cos x) \, dx$

(ii) $\int (\sin 5x - \cos 4x) \, dx$

(iii) $\int (\sin 5x + \sin 3x) \, dx.$

5. Determine

(i) $\int \sin \frac{1}{3}x \, dx$

(ii) $\int \cos \frac{1}{5}x \, dx$

(iii) $\int \left(\sin \frac{1}{2}x + \cos \frac{5}{7}x \right) dx.$

6. (i) $\int \sin^2 x \, dx$

(ii) $\int (1 + \sin^2 x) \, dx$

(iii) $\int (\cos^2 x - 1) \, dx.$

7. (i) $\int_{\pi/4}^{5\pi/4} \cos^2 x \, dx$

(ii) $\int_0^{3\pi/2} (\sin^2 x + 1) \, dx.$

8. (i) $\int \sin x \, d(\sin x)$

(ii) $\int \cos x \, d(\cos x)$

(iii) $\int \tan x \, d(\tan x).$

9. (i) $\int \tan^3 x \, d(\tan x)$

(ii) $\int \cot^4 x \, d(\cot x).$

10. (i) $\int \cos(3 - 5x) \, dx$

(ii) $\int \sin(3x - 4) \, dx.$

11. (i) $\int \tan^{\frac{1}{2}} \sqrt{x} \, d(\tan \sqrt{x})$

(ii) $\int \cot^{\frac{3}{2}} (\sqrt{x} - 1) \, d[\cot(\sqrt{x} - 1)].$

12. (i) $\int \sin^2 \frac{1}{4}x \, dx$

(ii) $\int \cos^2 \frac{1}{2}x \, dx.$

3

Exponential Functions

An exponential function is a function whose variable is the exponent, $a^x, 2^x, 3^x, e^x$ are exponential functions.

$\int e^x \, dx = e^x + k$ the derived function of $e^x + k$ is e^x

$\int e^{kx} \, dx = \dfrac{e^{kx}}{k} + c.$

To show this result, let $u = kx$, $\dfrac{du}{dx} = k$

$\int e^u \dfrac{du}{k} = \dfrac{e^u}{k} + c = \dfrac{e^{ku}}{k} + c$

therefore $\boxed{\int e^{kx} \, dx = \dfrac{e^{kx}}{k} + c}$

WORKED EXAMPLE 34

$\int e^{3x} \, dx = \dfrac{1}{3} e^{3x} + k.$

WORKED EXAMPLE 35

$\int e^{\frac{1}{2}x} \, dx = 2 e^{\frac{1}{2}x} + k.$

WORKED EXAMPLE 36

$\int 5 e^{-\frac{3}{5}x} \, dx = \dfrac{5 e^{-\frac{3}{5}x}}{-\frac{3}{5}} + k = -\dfrac{25}{3} e^{-\frac{3}{5}x} + k.$

To integrate the exponential function a^x where a is a different constant to e.

$\int a^x \, dx.$

Let $y = a^x$, taking logarithms to the base e, $\ln y = \ln a^x$

$\ln y = x \ln a$

$\dfrac{1}{y} \dfrac{dy}{dx} = \ln a \qquad \dfrac{dy}{dx} = y \ln a = a^x \ln a$

$\boxed{\int a^x \, dx = \dfrac{a^x}{\ln a} + c}$

WORKED EXAMPLE 37

$\int b^x \, dx = \dfrac{b^x}{\ln b} + k.$

WORKED EXAMPLE 38

$\int 3^x \, dx = \dfrac{3^x}{\ln 3} + k.$

• If the base is e, then $\ln e = 1$ and

$\int e^x \, dx = \dfrac{e^x}{\ln e} + k = e^x + k$ as before.

WORKED EXAMPLE 39

$\int \dfrac{e^x + e^{-x}}{2} \, dx = \int \left(\dfrac{e^x}{2} + \dfrac{e^{-x}}{2} \right) dx$

$= \dfrac{1}{2} e^x - \dfrac{e^{-x}}{2} + c.$

WORKED EXAMPLE 40

$\int \dfrac{e^x - e^{-x}}{2} \, dx = \dfrac{1}{2} \left(e^x - \dfrac{e^{-x}}{-1} \right) + k$

$= \dfrac{1}{2} (e^x + e^{-x}) + k.$

Worked Example 41

$$\int_0^2 e^{-x}\,dx = \left[\frac{e^{-x}}{-1}\right]_0^2 = \left[\frac{e^{-2}}{-1}\right] - \left[\frac{e^{-0}}{-1}\right].$$

$$= -\frac{1}{e^2} + 1 = 0.865 \text{ to 3 d.p.}$$

Worked Example 42

$$\int_3^4 e^{5x}\,dx = \left[\frac{e^{5x}}{5}\right]_3^4 = \frac{e^{20}}{5} - \frac{e^{15}}{5} = \frac{1}{5}(e^{20} - e^{15}).$$

Exercises 3

1. Find the indefinite integrals of the following functions:
   (i) $e^{-\frac{3x}{2}}$
   (ii) $e^{\frac{x}{5}}$
   (iii) $e^{-\frac{x}{4}}$
   (iv) $3e^{5x}$
   (v) $-5e^{-x}$.

2. Find the definite integrals
   (i) $\int_0^1 e^{-\frac{3x}{2}}\,dx$
   (ii) $\int_1^2 e^{\frac{x}{5}}\,dx$
   (iii) $\int_2^3 e^{-\frac{x}{4}}\,dx$
   (iv) $\int_3^4 3e^{5x}\,dx$ and
   (v) $\int_{-2}^{-1} -5e^{-x}\,dx.$

3. Evaluate the following definite integrals
   (i) $\int_0^{0.1} 2^x\,dx$
   (ii) $\int_{0.1}^{0.2} 3^x\,dx$
   (iii) $\int_1^2 5^{-x}\,dx$
   (iv) $\int_0^1 \frac{1}{e^x}\,dx$
   (v) $\int_3^5 \frac{1}{4^{-x}}\,dx.$

4. Show that $\int a^x\,dx = \frac{a^x}{\ln a} + c$ and hence deduce $\int e^x\,dx = e^x + c.$

5. Evaluate
   (i) $\int_{-1}^1 \frac{e^x + e^{-x}}{2}\,dx$
   (ii) $\int_{-1}^0 \frac{e^x - e^{-x}}{2}\,dx.$

6. Evaluate
   (i) $\int_0^1 \sqrt{e^z}\,dz$
   (ii) $\int_0^2 \sqrt[3]{e^z}\,dz$
   (iii) $\int_0^1 \sqrt[5]{e^{-t}}\,dt$
   (iv) $\int_2^3 \sqrt{e^{-y}}\,dy$
   (v) $\int_3^5 (e^x)^{\frac{1}{3}}\,dx$
   (vi) $\int_0^1 (e^{-x})^{\frac{2}{3}}\,dx.$

4

Logarithmic Functions

The Integral of $\frac{1}{x}$

The derivative of $\ln x$ is $\frac{1}{x}$. Therefore the integral of $\frac{1}{x}$ is $\ln x$

$$\int \frac{1}{x} \, dx = \ln|x| + k = \ln|x| + \ln|A| = \ln|Ax|.$$

For the logarithm to exit Ax must be a positive quantity. Therefore,

$$\ln x = \int_1^x \frac{1}{y} \, dy \quad \text{for } x > 0.$$

The curve $\frac{1}{y} = f(y)$ is rectangular hyperbola; if y tends to infinity, $\frac{1}{y} \to 0$ and when y tends to zero, $\frac{1}{y} \to \infty$.

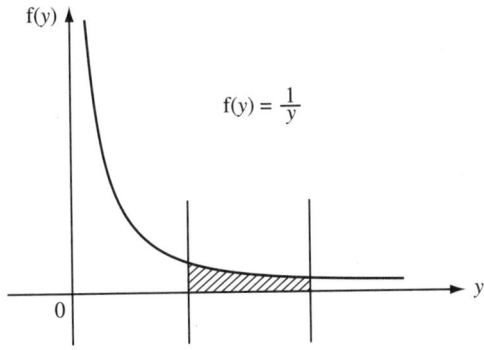

Fig. 7-I/12 The hatched area is equal to $\ln x$

$$\boxed{\int \frac{1}{x} \, dx = \ln|x| + k}$$

where $|x|$ denotes the positive quantity of x

$$\boxed{\int \frac{1}{ax+b} \, dx = \frac{\ln(ax+b)}{a} + k}$$

WORKED EXAMPLE 43

Integrate the following with respect to t.

(i) $\dfrac{1}{3t}$

(ii) $\dfrac{3}{t}$

(iii) $\dfrac{1}{-t+1}$

(iv) $\dfrac{3}{5t-1}$

(v) $\dfrac{1}{t+1}.$

Solution 43

(i) $\displaystyle\int \frac{1}{3t} \, dt = \frac{1}{3} \ln|t| + k$

(ii) $\displaystyle\int \frac{3}{t} \, dt = 3 \ln|t| + k$

(iii) $\displaystyle\int \frac{1}{-t+1} \, dt = \frac{\ln|1-t|}{-1} + k$

(iv) $\displaystyle\int \frac{3}{5t-1} \, dt = \frac{3 \ln|5t-1|}{5} + k$

(v) $\displaystyle\int \frac{1}{t+1} \, dt = \ln|1+t| + k.$

WORKED EXAMPLE 44

$$\int \frac{1}{(x-1)(x+2)} \, dx$$

we have to express $\dfrac{1}{(x-1)(x+2)}$ in partial fractions

$$\frac{1}{(x-1)(x+2)} \equiv \frac{A}{(x-1)} + \frac{B}{(x+2)}$$

$$1 \equiv A(x+2) + B(x-1)$$

If $x = -2$, $B = -\dfrac{1}{3}$ and if $x = 1$, $A = \dfrac{1}{3}$

$$\frac{1}{(x-1)(x+2)} = \frac{\frac{1}{3}}{x-1} - \frac{\frac{1}{3}}{x+2}$$

$$\int \left(\frac{\frac{1}{3}}{x-1} - \frac{\frac{1}{3}}{x+2} \right) dx$$

$$= \frac{1}{3} \ln|x-1| - \frac{1}{3} \ln|x+2| + k$$

$$= \frac{1}{3} \ln \left| \frac{x-1}{x+2} \right| + k = \frac{1}{3} \ln A \left| \frac{x-1}{x+2} \right|.$$

WORKED EXAMPLE 45

$$\int \frac{4x}{(1-2x)(1+3x)} \, dx$$

Expressing $\dfrac{1}{(1-2x)(1+3x)}$ into partial fractions

$$\frac{4x}{(1-2x)(1+3x)} \equiv \frac{A}{1-2x} + \frac{B}{1+3x}$$

$$4x \equiv A(1+3x) + B(1-2x)$$

If $x = -\dfrac{1}{3}$ \qquad\qquad If $x = \dfrac{1}{2}$

$$-\frac{4}{3} \equiv B\left(1 + \frac{2}{3}\right) \qquad 4\left(\frac{1}{2}\right) = A\left(1 + \frac{3}{2}\right)$$

$$-\frac{4}{3} = B\frac{5}{3} \qquad\qquad 2 = A\frac{5}{2}$$

$$B = -\frac{4}{5} \qquad\qquad A = \frac{4}{5}$$

$$\frac{4x}{(1-2x)(1+3x)} \equiv \frac{\frac{4}{5}}{1-2x} - \frac{\frac{4}{5}}{1+3x}$$

$$\int \left(\frac{\frac{4}{5}}{1-2x} - \frac{\frac{4}{5}}{1+3x} \right) dx$$

$$= \frac{4}{5} \frac{\ln(1-2x)}{(-2)} - \frac{4}{5} \frac{\ln(1+3x)}{3} + \ln A$$

$$= -\frac{2}{5} \ln(1-2x) - \frac{4}{15} \ln(1+3x) + \ln A$$

$$= \ln \frac{A}{(1-2x)^{\frac{2}{5}}(1+3x)^{\frac{4}{15}}}$$

Exercises 4

1. Integrate the following with respect to x

   (i) $\dfrac{2}{x}$

   (ii) $\dfrac{1}{2x}$

   (iii) $\dfrac{5}{1-x}$

   (iv) $\dfrac{1}{1+x}$

   (v) $\dfrac{7}{1-2x}$

   (vi) $\dfrac{1}{5x-1}$

   (vii) $\dfrac{1}{x(x-1)}$

   (viii) $\dfrac{1}{x(x+1)(x+2)}$

   (ix) $\dfrac{1}{(1-2x)(1-x)}$

   (x) $\dfrac{1}{(3x+5)(2x-1)}$.

   Each function should be written down with its integral sign and the differential dx before being integrated.

2. Evaluate the following integrals.

(i) $\int_1^2 \frac{1}{x}\,dx$

(ii) $\int_{-3}^{-2} \frac{1}{1-x}\,dx$

(iii) $\int_5^8 \frac{1}{x(x+1)}\,dx$

(iv) $\int_{0.1}^{0.3} \frac{1}{1-x}\,dx$

(v) $\int_0^{0.5} \frac{x}{(1+x)(1-x)}\,dx$

(vi) $\int_2^3 \frac{x^3}{(x+1)^2(x-1)^2}\,dx$.

3. $\int_{\frac{5}{2}}^3 \frac{1}{(2x-1)^2(x+3)^2}\,dx$.

4. Integrate the following, by writing each function with its integral sign and the differential before being integrated.

(i) $\frac{1}{5t}$

(ii) $\frac{1}{1+z}$

(iii) $\frac{1}{1-3u}$

(iv) $\frac{x}{1-x^2}$

(v) $\frac{y}{(y-1)(y+2)}$.

5. (i) $\int \frac{dx}{2x+5}$

(ii) $\int \frac{2dx}{3x-4}$

(iii) $\int \frac{2-x}{2x(1-x)}\,dx$

(iv) $\int \frac{5x^2-7x-9}{x^2}\,dx$.

6. (i) $\int \frac{1}{a+bx}\,dx$

(ii) $\int \frac{a}{a-bx}\,dx$

(iii) $\int \frac{1}{(x-1)(x+2)}\,dx$.

7. (i) $\int \frac{dx}{4-x^2}$

(ii) $\int \frac{dx}{x^2-25}$

(iii) $\int \frac{dx}{x(x-1)}$

(iv) $\int \frac{dx}{x^2-4x-5}$.

8. (i) $\int_0^{\frac{1}{\sqrt{3}}} \left(\frac{1}{1-u} + \frac{1}{1+u} \right) du$

(ii) $\int_0^1 \frac{y^2\,dy}{2-y^3}$

(iii) $\int_1^2 \frac{3x^5}{2x^6+1}\,dx$.

9. (i) $\int_2^5 \frac{x+3}{x-1}\,dx$

(ii) $\int_4^5 \frac{x+7}{x-3}\,dx$

(iii) $\int_{\frac{1}{2}}^{\frac{3}{4}} \frac{1+x}{1-x}\,dx$.

10. Determine the following indefinite integrals given that each rational function is decomposed, into partial fractions adjacently.

(i) $\int \dfrac{6(2-x)}{9-x^2}\,dx \quad \left(-\dfrac{1}{3-x}+\dfrac{5}{3+x}\right)$

(ii) $\int \dfrac{8x+3}{(x+3)(x-4)}\,dx \quad \left(\dfrac{3}{x+3}+\dfrac{5}{x-4}\right)$

(iii) $\int \dfrac{6-x}{x(x+3)}\,dx \quad \left(\dfrac{2}{x}-\dfrac{3}{x+3}\right)$

(iv) $\int \dfrac{2+x-x^2}{x^2(x+2)}\,dx \quad \left(\dfrac{1}{x^2}-\dfrac{1}{x+2}\right)$

(v) $\int \dfrac{6x^2+13x+3}{x(x+1)(x+3)}\,dx$

$\left(\dfrac{1}{x}+\dfrac{2}{x+1}+\dfrac{3}{x+3}\right)$

(vi) $\int \dfrac{x^3-4x^2-18x+23}{(x+2)(x-3)(x-5)}\,dx$

$\left(\dfrac{x+1}{x-3}+\dfrac{1}{x+2}-\dfrac{3}{x-5}\right)$

(vii) $\int \dfrac{x}{x+1}\,dx \quad \left(1-\dfrac{1}{x+1}\right)$

(viii) $\int \dfrac{2x+2}{x+2}\,dx \quad \left(1+\dfrac{x}{x+2}\right)$

(ix) $\int \dfrac{x^2+4x-3}{x(x-1)}\,dx \quad \left(\dfrac{x+1}{x-1}+\dfrac{3}{x}\right)$

(x) $\int \dfrac{3x^3+6x^2+6x+2}{(x+1)^3(2x+1)}\,dx$

$\left(\dfrac{1}{(x+1)^3}-\dfrac{1}{(x+1)^2}+\dfrac{1}{x+1}+\dfrac{1}{2x+1}\right)$

(xi) $\int \dfrac{x(x+1)}{(x^2+1)(x-1)}\,dx \quad \left(\dfrac{1}{x^2+1}+\dfrac{1}{x-1}\right)$

(xii) $\int \dfrac{2x^2-2x-2}{(2x-1)(2x^2-3)}\,dx$

$\left(\dfrac{1}{2x-1}-\dfrac{1}{2x^2-3}\right)$

(xiii) $\int \dfrac{x^3+2x^2-9x-19}{(x+2)^2(x^2-5)}\,dx$

$\left(\dfrac{1}{x+2}+\dfrac{1}{(x+2)^2}-\dfrac{1}{x^2-5}\right).$

5

Integration by Inspection

This type of integration is applicable for products or quotients and are called integrating products and integrating quotients respectively.

Certain types of functions can be integrated by observing that one function is the differential coefficient of another function. This is best illustrated by some examples.

WORKED EXAMPLE 46

$\int 3x^2 \left(x^3 + 1\right) dx$. It is observed that $3x^2$ is the differential coefficient of x^3+1, therefore $d(x^3+1) = 3x^2\, dx$.

or let $y = x^3 + 1$ differentiating with respect to x

$$\frac{dy}{dx} = 3x^2$$

or $\dfrac{d(x^3 + 1)}{dx} = 3x^2$

$$d\left(x^3 + 1\right) = 3x^2\, dx$$

$$\int 3x^2 \left(x^3 + 1\right) dx = \int \left(x^3 + 1\right) d\left(x^3 + 1\right)$$

the variable is now different, $x^3 + 1$ instead of x

$$\int \left(x^3 + 1\right) d\left(x^3 + 1\right) = \frac{\left(x^3 + 1\right)^2}{2} + c$$

therefore $\int 3x^2 \left(x^3 + 1\right) dx$

$$= \int \left(x^3 + 1\right) d\left(x^3 + 1\right)$$

$$= \frac{\left(x^3 + 1\right)^2}{2} + c.$$

WORKED EXAMPLE 47

$\int 6x\sqrt{3x^2 + 1}\, dx$. It is observed that the derivative of $3x^2 + 1$ is $6x$.

$$\int 6x\sqrt{3x^2 + 1}\, dx = \int \sqrt{3x^2 + 1}\, d\left(3x^2 + 1\right)$$

$$= \frac{\left(3x^2 + 1\right)^{\frac{1}{2}+1}}{\frac{1}{2} + 1} + c$$

$$= \frac{2}{3}\left(3x^2 + 1\right)^{\frac{3}{2}} + c.$$

WORKED EXAMPLE 48

$\int x\left(2x^2 - 1\right)^5 dx$. Obviously, we do not need to expand $(2x^2 - 1)^5$ and then multiply by x. The differential coefficient of $2x^2 - 1$ is $4x$.

$$\int \frac{4x}{4}\left(2x^2 - 1\right)^5 dx = \frac{1}{4}\int 4x\left(2x^2 - 1\right)^5 dx$$

$$= \frac{1}{4}\int \left(2x^2 - 1\right)^5 d\left(2x^2 - 1\right)$$

$$= \frac{1}{4}\frac{\left(2x^2 - 1\right)^6}{6}$$

$$= \frac{1}{24}\left(2x^2 - 1\right)^6 + c.$$

WORKED EXAMPLE 49

$\int \frac{\sin x}{\sqrt{\cos x}} dx$. It is observed that the differential coefficient of $\cos x$ is $-\sin x$

$$\int \frac{-d(\cos x)}{(\cos x)^{\frac{1}{2}}} = -\int (\cos x)^{-\frac{1}{2}} d(\cos x)$$

$$= -\frac{(\cos x)^{\frac{1}{2}}}{\frac{1}{2}} + c$$

$$= -2\sqrt{\cos x} + c.$$

All the above can be determined by making the appropriate substitution, consider again worked example 47.

$\int 3x^2 (x^3 + 1) dx$. Let $u = x^3 + 1$

$$\frac{du}{dx} = 3x^2 \quad \text{or} \quad dx = \frac{du}{3x^2}$$

$$\int 3x^2 u \frac{du}{3x^2} = \int u \, du = \frac{u^2}{2} + c$$

$$s = \frac{(x^3 + 1)^2}{2} + c$$

the procedure is lengthier.

WORKED EXAMPLE 50

$$\int \tan x \, dx = \int \frac{\sin x}{\cos x} dx = \int \frac{-d(\cos x)}{\cos x}$$

$$= -\ln|\cos x| + c$$

$$= \ln|\sec x| + c.$$

WORKED EXAMPLE 51

$$\int \cot x \, dx = \int \frac{\cos x}{\sin x} dx = \int \frac{d(\sin x)}{\sin x}$$

$$= \ln|\sin x| + c$$

$$= \ln|\sin x| + c.$$

Let us try again the appropriate substitution for the last example.

$$\int \cot x \, dx = \int \frac{\cos x}{\sin x} dx.$$

Let $\sin x = u$, $\frac{du}{dx} = \cos x$ or $dx = \frac{du}{\cos x}$

$$\int \cot x \, dx = \int \frac{\cos x}{\sin x} dx = \int \frac{\cos x}{u} \cdot \frac{du}{\cos x}$$

$$= \int \frac{du}{u} = \ln|u| + c = \ln|\sin x| + c.$$

Obviously there is a difficulty which function is to be substituted $\sin x$ or $\cos x$

$\int \frac{\cos x}{\sin x} dx$. If $u = \cos x$, $\frac{du}{dx} = -\sin x$ or

$$dx = \frac{du}{-\sin x}.$$

$-\int \frac{u}{\sin x} \cdot \frac{du}{\sin x}$ the integral becomes very complicated and cannot be solved, therefore the suitable substitution is $u = \sin x$ as shown above. The inspection method is the best in solving such integrals.

Let us now consider

WORKED EXAMPLE 52

$\int \frac{1}{x} \ln x \, dx$. It is observed that the differential coefficient of $\ln x$ is $\frac{1}{x}$, therefore

$$\int \frac{1}{x} \ln x \, dx = \int \ln x \, d(\ln x) = \frac{1}{2} (\ln x)^2 + c.$$

There is another line of thought we would like to change the variable x to another suitable variable, what is the new variable such that if it is differentiated will give either $\frac{1}{x}$ or $\ln x$, we do not readily know the function whose derivative is $\ln x$ therefore the suitable new variable we are looking for is $\ln x$ whose derivative is $\frac{1}{x}$ hence $\frac{1}{x} dx = d(\ln x)$.

WORKED EXAMPLE 53

$$\int 2x e^{x^2} dx = \int e^{x^2} d(x^2)$$

$$= e^{x^2} + c \text{ where } 2x \, dx = d(x^2).$$

Worked Example 54

$$\int \frac{\sin x}{\cos^6 x} dx = -\int \frac{d(\cos x)}{\cos^6 x}$$

$$= -\int (\cos x)^{-6} d(\cos x)$$

$$= -\frac{(\cos x)^{-5}}{-5} + c = \frac{1}{5\cos^5 x} + c.$$

Worked Example 55

$$\int \frac{x\, dx}{\sqrt{7x^2+1}} = \frac{1}{14} \int \frac{d(7x^2+1)}{\sqrt{7x^2+1}}$$

$$= \frac{1}{14} \frac{(7x^2+1)^{\frac{1}{2}}}{\frac{1}{2}} + c$$

$$= \frac{1}{7}\left(7x^2+1\right)^{\frac{1}{2}} + c$$

Exercises 5

1. Integrate the following with respect to x by inspecting that one function is the derivative of another.

   (i) $x^2\sqrt{x^3+1}$

   (ii) $\dfrac{x^2}{\sqrt{x^3+1}}$

   (iii) $x\left(x^2+1\right)^7$

   (iv) $\dfrac{\sec^2 x}{\tan x}$

   (v) $\cosec^2 x \cot x$

   (vi) $\sec^2 x \tan x$

   (vii) $\sec x \tan x \sec x$

   (viii) $-\cosec^2 x \cot x$

   (ix) $\tan x \sec^2 x$

   (x) $\dfrac{(\sec x \tan x)}{\sec x}$

   (xi) $\dfrac{-\cosec x \cot x}{\cosec x}$

   (xii) $\dfrac{\cos x}{\sin^3 x}$.

2. Evaluate the following products:

   (i) $\displaystyle\int_0^{\pi/4} \cos^2 x \sin^3 x \, dx$

   (ii) $\displaystyle\int_{\pi/6}^{\pi/2} \cos^{10} x \sin x \, dx$

   (iii) $\displaystyle\int_0^{\pi} \cos^{15} x \sin^5 x \, dx$

   (iv) $\displaystyle\int_{\pi/4}^{\pi/2} \cosec^2 x \cos x \, dx$

   (v) $\displaystyle\int_0^{2\pi} \sec^2 x \sin x \, dx$

   (vi) $\displaystyle\int_0^{\pi/2} \cos^n x \sin^3 x \, dx$

   (vii) $\displaystyle\int_0^{\pi/2} \cos^3 x \sin^2 x \, dx$

   (viii) $\displaystyle\int_0^{\pi/2} \cos^5 x \sin^n x \, dx$

   (ix) $\displaystyle\int_{-\pi/2}^{\pi/2} \sin^9 x \cos^3 x \, dx$

   (x) $\displaystyle\int_{-\pi}^{\pi} \cos^n x \sin x \, dx$

(xi) $\int_0^{\pi/4} \sin^{11} x \cos^5 x \, dx$

(xii) $\int_0^{\pi/2} \sin^n x \, d(\sin x)$.

3. (i) $\int (3 + e^x)^{\frac{5}{2}} \, d(3 + e^x)$

(ii) $\int \sin^{\frac{7}{2}} x \, d(\sin x)$

(iii) $\int \cos^{\frac{9}{2}} x \, d(\cos x)$

(iv) $\int \tan^{\frac{5}{2}} x \, d(\tan x)$

(v) $\int \csc^{\frac{1}{2}} x \, d(\csc x)$

(vi) $\int \sec^{\frac{3}{2}} x \, d(\sec x)$

(vii) $\int (1 + x^m)^n \, d(1 + x^m)$

(viii) $\int (1 + x^3)^5 \, d(1 + x^3)$

(ix) $\int (1 + \cos\theta)^{\frac{7}{2}} \, d(1 + \cos\theta)$

(x) $\int e^{x^3+1} \, d(x^3 + 1)$

(xi) $\int (x^3 + 3x^2 + 5x - 7)^7 \, d(x^3 + 3x^2 + 5x - 7)$

(xii) $\int [\ln(x^2 + 3x - 1)]^5 \, d[\ln(x^2 + 3x - 1)]$

(xiii) $\int \sin^2 x \, d(\sin x)$

(xiv) $\int (\cos x)^{-2} \, d(\cos x)$

(xv) $\int (\cos x^{-1})^3 \, d(\cos x^{-1})$.

4. (i) $\int \csc^2 x \sqrt{\tan x} \, dx$

(ii) $\int \dfrac{\sec^2 x}{\sqrt{\tan x}} \, dx$

(iii) $\int \cos^5 3x \, dx$

(iv) $\int \sin x \cos^2 x \, dx$

(v) $\int \sin 3x \cos^3 3x \, dx$

(vi) $\int \cos^5 2x \, dx$

(vii) $\int \cos^2 \dfrac{x}{2} \, dx$

(viii) $\int \sin^2 \dfrac{x}{4} \, dx$

(ix) $\int 4x^3 \sqrt{x^4 - 1} \, dx$

(x) $\int \dfrac{2x}{(x^2 - 1)^5} \, dx$

(xi) $\int 2x \, e^{x^2} \, dx$

(xii) $\int \dfrac{1}{x} \ln x \, dx$

(xiii) $\int \dfrac{\ln 7x}{x} \, dx$

(xiv) $\int 3x \sqrt{x^2 - 5} \, dx$

(xv) $\int \cos^5 2x \, dx$

(xvi) $\int \sec^2 x \tan^5 x \, dx$

(xvii) $\int \csc^2 x \cot^3 x \, dx$

(xviii) $\int \dfrac{\sin x}{\sqrt{\cos x}} \, dx$

(xix) $\int \cos x \sqrt{\sin x} \, dx$

(xx) $\int 4x^3 e^{x^4} \, dx$.

6

Hyperbolic Functions

Definitions

$\sinh x = \frac{1}{2}(e^x - e^{-x})$

$\cosh x = \frac{1}{2}(e^x + e^{-x})$

where x is expressed in radians or numbers, <u>not</u> in degrees. These functions are called hyperbolic functions since the parametric equations $x = a\cosh u$, $y = b\sinh u$, verify the rectangular hyperbola.

$\frac{x^2}{a^2} - \frac{y^2}{b^2} = 1$

$\frac{x^2}{a^2} - \frac{y^2}{b^2} = \frac{a^2 \cosh^2 u}{a^2} - \frac{b^2 \sinh^2 u}{b^2}$

$= \cosh^2 u - \sinh^2 u = 1$

$\cosh^2 x - \sinh^2 x = \frac{1}{4}(e^x + e^{-x})^2 - \frac{1}{4}(e^x - e^{-x})^2$

$= \frac{1}{4}e^{2x} + \frac{1}{2} + \frac{1}{4}e^{-2x}$

$ - \frac{1}{4}e^{2x} + \frac{1}{2} - \frac{1}{4}e^{-2x}$

$\boxed{\cosh^2 x - \sinh^2 x \equiv 1}$

WORKED EXAMPLE 56

Using the definitions

$\cosh x = \frac{1}{2}(e^x + e^{-x})$

$\sinh x = \frac{1}{2}(e^x - e^{-x})$,

show the following identities:

(i) $\cosh(x+y)$
$ = \cosh x \cosh y + \sinh x \sinh y$

(ii) $\sinh 3x = 3 \sinh x + 4 \sinh^3 x$

(iii) $\cosh 3x = 4 \cosh^3 x - 3 \cosh x$.

Solution 56

(i) $\cosh(x+y)$

$= \frac{1}{2}(e^{x+y} + e^{-(x+y)})$

$\cosh x \cosh y + \sinh x \sinh y$

$= \frac{1}{2}(e^x + e^{-x}) \frac{1}{2}(e^y + e^{-y})$

$ + \frac{1}{2}(e^x - e^{-x}) \frac{1}{2}(e^y - e^{-y})$

$= \frac{1}{4}(e^{x+y} + e^{x-y} + e^{-x+y} + e^{-x-y}$

$\phantom{= \frac{1}{4}(} + e^{x+y} - e^{x-y} - e^{-x+y} + e^{-x-y})$

$= \frac{1}{4}(2e^{x+y} + 2e^{-(x+y)})$

$= \frac{1}{2}(e^{x+y} + e^{-(x+y)})$

therefore

$\boxed{\cosh(x+y) = \cosh x \cosh y + \sinh x \sinh y}$

(ii) $\sinh 3x = 3 \sinh x + 4 \sinh^3 x$

$\sinh 3x = \frac{1}{2}(e^{3x} - e^{-3x})$

29

$3\sinh x + 4\sinh^3 x$

$= \frac{3}{2}(e^x - e^{-x}) + 4 \cdot \frac{1}{8}(e^x - e^{-x})^3$

$= \frac{3}{2}e^x - \frac{3}{2}e^{-x} + \frac{1}{2}e^{3x}$

$\quad - \frac{3}{2}e^x + \frac{3}{2}e^{-x} - \frac{1}{2}e^{-3x}$

$= \frac{1}{2}(e^{3x} - e^{-3x})$

therefore

$\boxed{\sinh 3x = 3\sinh x + 4\sinh^3 x}$

(iii) $\cosh 3x = 4\cosh^3 x - 3\cosh x$.

$\cosh 3x = \frac{1}{2}(e^{3x} + e^{-3x})$

$4\cosh^3 x - 3\cosh x$

$= \frac{4}{8}(e^x + e^{-x})^3 - \frac{3}{2}(e^x + e^{-x})$

$= \frac{1}{2}(e^{3x} + 3e^x + 3e^{-x} + e^{-3x})$

$\quad - \frac{3}{2}e^x - \frac{3}{2}e^{-x}$

$= \frac{1}{2}(e^{3x} + e^{-3x})$.

Trigonometric and Hyperbolic Identities

Replace $\sin x$ by $i \sinh x$ in the following trigonometric identities where $i = \sqrt{-1}$. Osborne's rule states that $\sin^2 x$ is replaced by $-\sinh^2 x$.

Trigonometric

$\sin^2 x + \cos^2 x = 1$

$1 + \tan^2 x = \sec^2 x$

$1 + \cot^2 x = \operatorname{cosec}^2 x$

$\cos 2x = 2\cos^2 x - 1$

$\cos 2x = 1 - 2\sin^2 x$

$\cos x = 2\cos^2 \frac{x}{2} - 1$

$\cos x = 1 - 2\sin^2 \frac{x}{2}$

$\tan 2x = \dfrac{2\tan x}{1 - \tan^2 x}$

$\sin(x \pm y) = \sin x \cos y \pm \sin y \cos x$

$\cos(x \pm y) = \cos x \cos y \mp \sin x \sin y$

$\tan(x \pm y) = \dfrac{\tan x \pm \tan y}{1 \mp \tan x \tan y}$

$2\sin x \cos y = \sin(x+y) + \sin(x-y)$

$2\cos x \cos y = \cos(x+y) + \cos(x-y)$

$2\sin x \sin y = \cos(x-y) - \cos(x+y)$

$\sin x + \sin y = 2\sin\dfrac{x+y}{2}\cos\dfrac{x-y}{2}$

$\sin x - \sin y = 2\cos\dfrac{x+y}{2}\sin\dfrac{x-y}{2}$

$\cos x + \cos y = 2\cos\dfrac{x+y}{2}\cos\dfrac{x-y}{2}$

$\cos x - \cos y = -2\sin\dfrac{x+y}{2}\sin\dfrac{x-y}{2}$

Hyperbolic

$\cosh^2 x - \sinh^2 x = 1$

$1 - \tanh^2 x = \operatorname{sech}^2 x$

$1 - \coth^2 x = -\operatorname{cosech}^2 x$

$\cosh 2x = 2\cosh^2 x - 1$

$\cosh 2x = 1 + 2\sinh^2 x$

$\cosh x = 2\cosh^2 \frac{x}{2} - 1$

$\cosh x = 1 + 2\sinh^2 \frac{x}{2}$

$\tanh 2x = \dfrac{2\tanh x}{1 + \tanh^2 x}$

$\sinh(x \pm y) = \sinh x \cosh y \pm \sinh y \cosh x$

$\cosh(x \pm y) = \cosh x \cosh y \pm \sinh x \sinh y$

$$\tanh(x \pm y) = \frac{(\tanh x \pm \tanh y)}{1 \pm \tanh x \tanh y}$$

$$2 \sinh x \cosh y = \sinh(x+y) + \sinh(x-y)$$

$$2 \cosh x \cosh y = \cosh(x+y) + \cosh(x-y)$$

$$-2 \sinh x \sinh y = \cosh(x-y) - \cosh(x+y)$$

$$\sinh x + \sinh y = 2 \sinh \frac{x+y}{2} \cosh \frac{x-y}{2}$$

$$\sinh x - \sinh y = 2 \cosh \frac{x+y}{2} \sinh \frac{x-y}{2}$$

$$\cosh x + \cosh y = 2 \cosh \frac{x+y}{2} \cosh \frac{x-y}{2}$$

$$\cosh x - \cosh y = 2 \sinh \frac{x+y}{2} \sinh \frac{x-y}{2}.$$

Graphs of Hyperbolic Functions

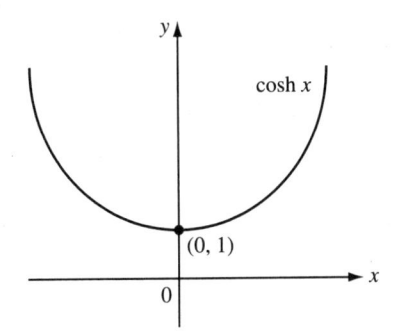

Fig. 7-I/13

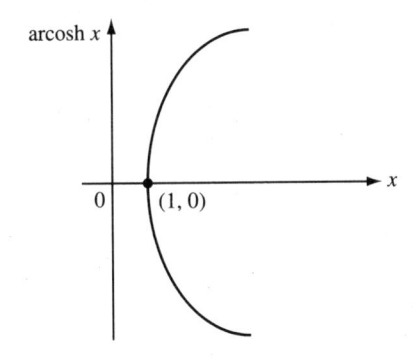

Fig. 7-I/14

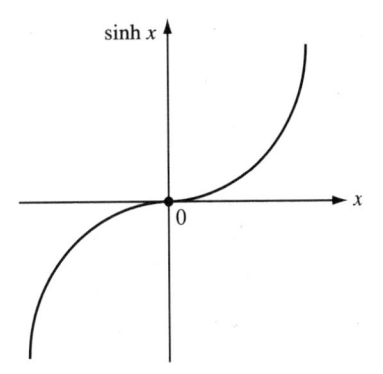

Fig. 7-I/15

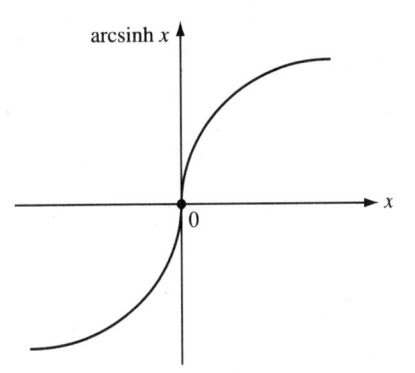

Fig. 7-I/16

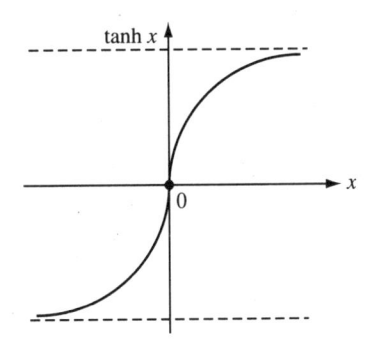

Fig. 7-I/17

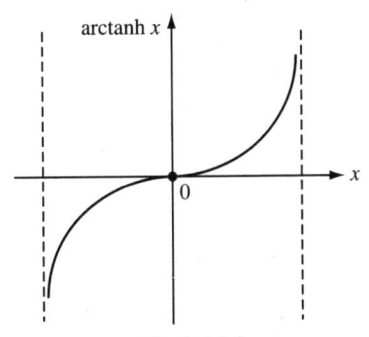

Fig. 7-I/18

Inverse Hyperbolic Functions

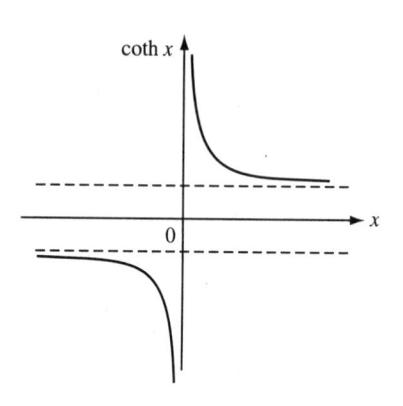

Fig. 7-I/19 — coth x

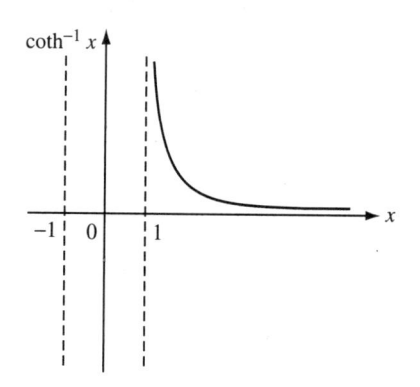

Fig. 7-I/20 — $\coth^{-1} x$

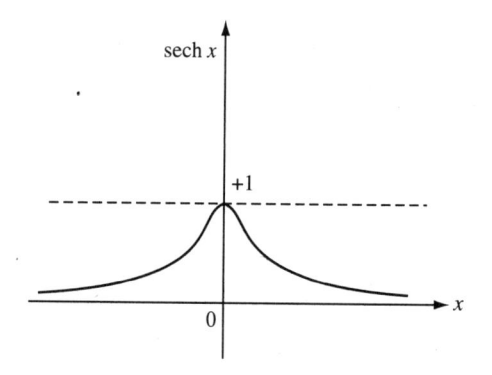

Fig. 7-I/21 — sech x

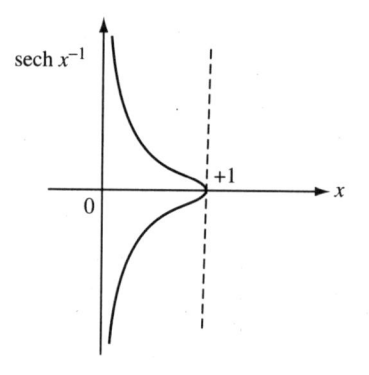

Fig. 7-I/22 — sech x^{-1}

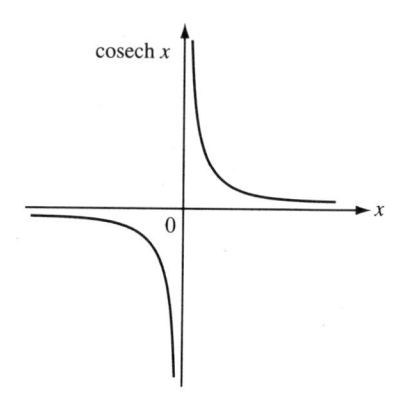

Fig. 7-I/23 — cosech x

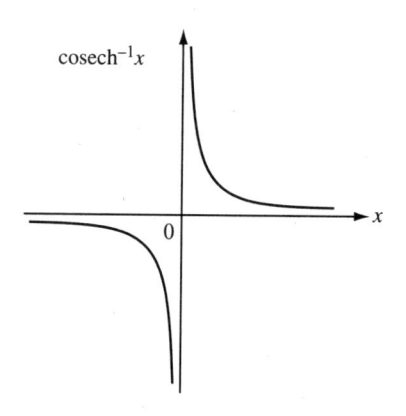

Fig. 7-I/24 — $\operatorname{cosech}^{-1} x$

$$\sinh^{-1} x = \ln\left[x + \sqrt{x^2 + 1}\right]$$

$$\cosh^{-1} x = \ln\left[x \pm \sqrt{x^2 - 1}\right] \; x \geq 1$$

$$\tanh^{-1} x = \frac{1}{2} \ln\left(\frac{1 + x}{1 - x}\right)$$

$$\operatorname{cosech}^{-1} x = \ln\left[\frac{1}{x} + \sqrt{\frac{1+x^2}{x^2}}\right] \quad |x| > 0$$

$$\operatorname{sech}^{-1} x = \ln\left[\frac{1}{x} \pm \sqrt{\frac{1-x^2}{x^2}}\right] \quad 0 < x \le 1$$

$$\operatorname{coth}^{-1} x = \frac{1}{2}\ln\left(\frac{x+1}{x-1}\right) \quad |x| > 1.$$

Inverse Hyperbolic Functions

$$y = \sinh^{-1} x, \; x = \sinh y = \frac{1}{2}\left(e^y - e^{-y}\right)$$

$e^y - e^{-y} = 2x$, multiplying by e^y each term

$$e^{2y} - 2xe^y - 1 = 0$$

let $W = e^y$

$$W^2 - 2xW - 1 = 0$$

$$W = \frac{2x \pm \sqrt{4x^2 + 4}}{2} = x \pm \sqrt{x^2 + 1}$$

$$e^y = x \pm \sqrt{x^2 + 1}$$

taking logarithms to the base e

$y = \ln\left(x + \sqrt{x^2 + 1}\right)$, $\ln\left(x - \sqrt{x^2 + 1}\right)$ is not defined for all values of x

$$\boxed{\sinh^{-1} x = \ln\left(x + \sqrt{x^2 + 1}\right)}$$

$$y = \cosh^{-1} x, \; x = \cosh y = \frac{1}{2}\left(e^y + e^{-y}\right)$$

$e^y + e^{-y} = 2x$, multiplying each term by e^y, $e^{2y} - 2xe^y + 1 = 0$

$$e^y = \frac{2x \pm \sqrt{4x^2 - 4}}{2} = x \pm \sqrt{x^2 - 1}$$

$$y = \ln\left(x \pm \sqrt{x^2 - 1}\right)$$

$$y = \ln\left(x + \sqrt{x^2 - 1}\right) \text{ or}$$

$$y = \ln\left(x - \sqrt{x^2 - 1}\right)$$

$$\frac{1}{x + \sqrt{x^2 - 1}} = x - \sqrt{x^2 - 1}$$

$$1 = \left(x - \sqrt{x^2 - 1}\right)\left(x + \sqrt{x^2 - 1}\right)$$

$$= x^2 - (x^2 - 1)$$

therefore

$$y = \ln\left(x + \sqrt{x^2 - 1}\right) \quad \text{or} \quad y = \ln\frac{1}{x + \sqrt{x^2 - 1}}$$

$$y = -\ln\left(x + \sqrt{x^2 - 1}\right)$$

$$y = \pm\ln\left(x + \sqrt{x^2 - 1}\right)$$

$$\boxed{\cosh^{-1} x = \pm\ln\left(x + \sqrt{x^2 - 1}\right)} \; (x \ge 1)$$

the principal value of $\cosh^{-1} x$ is given by $\cosh^{-1} x = \ln\left(x + \sqrt{x^2 - 1}\right)$

$$y = \tanh^{-1} x$$

$$x = \tanh y = \frac{e^y - e^{-y}}{e^y + e^{-y}} \times \frac{e^y}{e^y}$$

$$x = \frac{e^{2y} - 1}{e^{2y} + 1}$$

$$\left(e^{2y} + 1\right) x = e^{2y} - 1$$

$$1 + x = e^{2y} - xe^{2y}$$

$$1 + x = e^{2y}(1 - x)$$

taking logarithms on both sides

$$2y = \ln\frac{1+x}{1-x}$$

$$\boxed{\tanh^{-1} x = \frac{1}{2}\ln\frac{1+x}{1-x}} \; (|x| < 1)$$

$$y = \operatorname{cosech}^{-1} x$$

$$x = \operatorname{cosech} y = \frac{2}{e^y - e^{-y}}$$

$$xe^y - xe^{-y} = 2$$

multiplying each term by e^y

$$xe^{2y} - 2e^y - x = 0$$

$$e^y = \frac{2 \pm \sqrt{4+4x^2}}{2x}$$

$$e^y = \frac{1 \pm \sqrt{(1+x^2)}}{x}$$

$$y = \ln\left(\frac{1 \pm \sqrt{(1+x^2)}}{x}\right)$$

$$\boxed{\operatorname{cosech}^{-1} x = \ln\left(\frac{1}{x} \pm \sqrt{\frac{1+x^2}{x^2}}\right)} \quad |x| > 0$$

$$y = \operatorname{sech}^{-1} x$$

$$x = \operatorname{sech} y = \frac{2}{e^y + e^{-y}}$$

$$xe^y + xe^{-y} = 2$$

$$xe^{2y} - 2e^y + x = 0$$

$$e^y = \frac{2 \pm \sqrt{(4-4x^2)}}{2x}$$

$$e^y = \frac{1 \pm \sqrt{(1-x^2)}}{x}$$

$$y = \ln\left(\frac{1}{x} \pm \sqrt{\left(\frac{1-x^2}{x^2}\right)}\right)$$

$$\boxed{\operatorname{sech}^{-1} x = \ln\left(\frac{1}{x} \pm \sqrt{\left(\frac{1-x^2}{x^2}\right)}\right)} \quad 0 < x \leq 1$$

$$y = \coth^{-1} x$$

$$x = \coth y = \frac{e^y + e^{-y}}{e^y - e^{-y}}$$

$$e^y x - e^{-y} x = e^y + e^{-y}$$

$$e^y x - e^y = xe^{-y} + e^{-y}$$

$$e^{2y} x - e^{2y} = x + 1$$

$$e^{2y} = \frac{x+1}{x-1}$$

$$2y = \ln \frac{x+1}{x-1}$$

$$y = \frac{1}{2} \ln \frac{x+1}{x-1}$$

$$\boxed{\coth^{-1} x = \frac{1}{2} \ln \frac{x+1}{x-1}} \quad |x| > 1.$$

Hyperbolic Functions

$f(x)$	$\int f(x)\,dx$
$\sinh x$	$\cosh x$
$\cosh x$	$\sinh x$
$\tanh x$	$\ln\lvert\cosh x\rvert$
$\coth x$	$\ln\lvert\sinh x\rvert$
$\operatorname{sech}^2 x$	$\tanh x$
$\operatorname{cosech}^2 x$	$-\coth x$
$\dfrac{1}{\sqrt{x^2+a^2}}$	$\sinh^{-1}\dfrac{x}{a}$
$\dfrac{1}{\sqrt{x^2-a^2}}$	$\cosh^{-1}\dfrac{x}{a}\,(x>a)$
$\dfrac{1}{a^2-x^2}$	$\dfrac{1}{a}\tanh^{-1}\dfrac{x}{a},\,(\lvert x\rvert < a)$
$\dfrac{1}{x^2-a^2}$	$-\dfrac{1}{a}\coth^{-1}\dfrac{x}{a}\,(\lvert x\rvert > a)$
$\operatorname{sech} x$	$\sin^{-1}(\tanh x)$

Indefinite Integrals of Hyperbolic Functions

$$\int \sinh x\,dx = \frac{1}{2}\int (e^x - e^{-x})\,dx$$

$$= \frac{1}{2}(e^x + e^{-x}) + c = \cosh x + c.$$

$$\int \cosh x\,dx = \frac{1}{2}\int (e^x + e^{-x})\,dx$$

$$= \frac{1}{2}(e^x - e^{-x}) + c = \sinh x + c.$$

$$\int \tanh x \, dx = \int \frac{\sinh x}{\cosh x} \, dx$$

$$= \int \frac{d(\cosh x)}{\cosh x}$$

$$= \ln \cosh x + c.$$

$$\int \coth x \, dx = \int \frac{\cosh x}{\sinh x} \, dx$$

$$= \int \frac{d(\sinh x)}{\sinh x}$$

$$= \ln \sinh x + c.$$

$$\int \operatorname{sech} x \, dx = \int \frac{2}{e^x + e^{-x}} \, dx \quad \text{let } u = e^x \quad \frac{du}{dx} = e^x$$

$$= \int \frac{2 \, du}{e^x (e^x + e^{-x})} = \int \frac{2 \, du}{u \left(u + \frac{1}{u} \right)}$$

$$= \int \frac{2 \, du}{u^2 + 1}$$

$$= 2 \tan^{-1} u = 2 \tan^{-1} e^x + c.$$

$$\int \operatorname{cosech} x \, dx = \int \frac{2}{e^x - e^{-x}} \, dx$$

$$= \int \frac{2 \, du}{e^x (e^x - e^{-x})} = \int \frac{2 \, du}{u \left(u - \frac{1}{u} \right)}$$

$$= \int \frac{2 \, du}{u^2 - 1} \quad \text{where } u = e^x, \, \frac{dy}{dx} = e^x$$

$$\frac{2}{u^2 - 1} \equiv \frac{A}{u - 1} + \frac{B}{u + 1} \Rightarrow 2$$

$$\equiv A(u + 1) + B(u - 1).$$

If $u = 1$, $A = 1$, and if $u = -1$, $B = -1$

$$\int \operatorname{cosech} x \, dx = \int \frac{2 \, du}{u^2 - 1} = \int \left(\frac{1}{u - 1} - \frac{1}{u + 1} \right) du$$

$$= \ln(u - 1) - \ln(u + 1) = \ln \frac{u - 1}{u + 1}$$

$$= \ln \frac{e^x - 1}{e^x + 1} + c$$

Hyperbolic Functions — 35

$$\int \sinh kx \, dx = \frac{1}{k} \cosh kx + c$$

$$\int \cosh kx \, dx = \frac{1}{k} \sinh kx + c$$

$$\int \tanh kx \, dx = \frac{1}{k} \ln \cosh kx + c$$

$$\int \coth kx \, dx = \frac{1}{k} \ln \sinh kx + c$$

$$\int \operatorname{sech} kx \, dx = \frac{1}{k} \tan^{-1} e^{kx} + c$$

$$\int \operatorname{cosech} kx \, dx = \frac{1}{k} \ln \frac{(e^{kx} - 1)}{(e^{kx} + 1)} + c.$$

WORKED EXAMPLE 57

Evaluate the following definite integrals:

(i) $\displaystyle\int_1^2 \tanh \frac{1}{2} x \, dx$

(ii) $\displaystyle\int_2^3 \operatorname{sech} \frac{3}{4} x \, dx$

(iii) $\displaystyle\int_1^3 \operatorname{cosech} 2x \, dx.$

Solution 57

(i) $\displaystyle\int_1^2 \tanh \frac{1}{2} x \, dx = \left[2 \ln \cosh \frac{1}{2} x \right]_1^2$

$$= 2 \ln \cosh 1 - 2 \ln \cosh \frac{1}{2}$$

$$= 0.86756 - 0.24022 = 0.627$$

(ii) $\displaystyle\int_2^3 \operatorname{sech} \frac{3}{4} x \, dx = \left[\frac{4}{3} \tan^{-1} e^{\frac{3x}{4}} \right]_2^3$

$$= \frac{4}{3} \tan^{-1} e^{2.25} - \frac{4}{3} \tan^{-1} e^{\frac{3}{2}}$$

$$= 1.954 - 1.8017 = 0.152$$

(iii) $\int_1^3 \text{cosech } 2x \, dx = \frac{1}{2}\left[\ln\frac{e^{2x}-1}{e^{2x}+1}\right]_1^3$

$= \frac{1}{2}\ln\frac{e^6-1}{e^6+1} - \frac{1}{2}\ln\frac{e^2-1}{e^2+1}$

$= -2.479 \times 10^{-3} + 0.136$

$= 0.134$

$\int \frac{1}{\sqrt{x^2-a^2}} dx \qquad \cosh^2 y - \sinh^2 y = 1$

let $x = a \cosh y \qquad \cosh^2 y - 1 = \sinh^2 y$

$\frac{dx}{dy} = a \sinh y \qquad \cosh^2 y = 1 + \sinh^2 y$

$\int \frac{a \sinh y \, dy}{\sqrt{a^2 \cosh^2 y - a^2}} = y = \cosh^{-1}\frac{x}{a}$

$\boxed{\int \frac{1}{\sqrt{x^2-a^2}} dx = \cosh^{-1}\frac{x}{a} + c \\ = \text{arcosh}\frac{x}{a} + c}$

$\int \frac{1}{\sqrt{x^2+a^2}} dx$

let $x = a \sinh y$

$\frac{dx}{dy} = a \cosh y$

$\int \frac{1}{\sqrt{a^2 \sinh^2 y + a^2}} a \cosh y \, dy = y$

$= \sinh^{-1}\frac{x}{a}$

$\boxed{\int \frac{1}{\sqrt{x^2+a^2}} dx = \sinh^{-1}\frac{x}{a} + c \\ = \text{arcsinh}\frac{x}{a} + c}$

WORKED EXAMPLE 58

Show, by using a suitable substitution that

$\int \frac{1}{\sqrt{9x^2-16}} dx = \frac{1}{3} \text{arcosh}\frac{3x}{4}.$

Solution 58

$\int \frac{1}{\sqrt{9x^2-16}} dx$

let $x = \frac{4}{3}\cosh y,$

$\frac{dx}{dy} = \frac{4}{3}\sinh y$

$\cosh^2 y - \sinh^2 y = 1$

$= \int \frac{1}{\sqrt{9\left(\frac{16}{9}\right)\cosh^2 y - 16}} \cdot \frac{4}{3}\sinh y \, dy$

$= \frac{1}{4}\int \frac{4}{3} \cdot \frac{\sinh y}{\sqrt{\cosh^2 y - 1}} dy$

$= \frac{1}{3}y$

$= \frac{1}{3}\text{arcosh}\frac{3x}{4}.$

The Integration of the Squares of the Hyperbolic Function

$\int \cosh^2 x \, dx = \int \frac{\cosh 2x + 1}{2} dx$

$= \frac{\sinh 2x}{4} + \frac{1}{2}x + c$

$\cosh 2x = 2\cosh^2 x - 1$

$\cosh^2 x = \frac{\cosh 2x + 1}{2}$

$\int \sinh^2 x \, dx = \int \frac{\cosh 2x - 1}{2} dx$

$= \frac{\sinh 2x}{4} - \frac{1}{2}x + c$

$\cosh 2x = 1 + 2\sinh^2 x$

$\sinh^2 x = \frac{\cosh 2x - 1}{2}$

$$\int \tanh^2 x \, dx = \int \left(1 - \text{sech}^2 x\right) dx$$

$$= x - \tanh x + c$$

$$1 - \tanh^2 x = \text{sech}^2 x$$

$$\tanh^2 x = 1 - \text{sech}^2 x$$

$$\int \coth^2 x \, dx = \int (1 + \text{cosech}^2 x) \, dx$$

$$= x - \coth x + c$$

$$1 - \coth^2 x = -\text{cosech}^2 x$$

$$1 + \text{cosech}^2 x = \coth^2 x$$

$$\int \text{sech}^2 x \, dx = \tanh x + c$$

$$\int \text{cosech}^2 x \, dx = -\coth x + c.$$

The Integration of the Product of Hyperbolic Functions

$\int \sinh kx \cosh nx \, dx$ where k and n are constants
$\sinh(x+y) = \sinh x \cosh y + \sinh y \cosh x$ ...(1)
using the rule, to replace $\sin x$ by $i \sinh x$ in the identity $\sin(x+y) = \sin x \cos y + \sin y \cos x$

that is, $\sinh(x+y) = \sinh x \cosh y + \sinh y \cosh x$
...(1)

similarly
$\sinh(x-y) = \sinh x \cosh y - \sinh y \cosh x$ ...(2)

Adding (1) and (2)

$$\sinh(x+y) + \sinh(x-y) = 2 \sinh x \cosh y$$

$$\sinh x \cosh y = \frac{1}{2}[\sinh(x+y) + \sinh(x-y)]$$

$$\int \sinh kx \cosh nx \, dx$$

$$= \frac{1}{2} \int [\sinh(k-n)x + \sinh(k+n)x] \, dx$$

$$= \frac{1}{2} \frac{\cosh(k-n)x}{k-n} + \frac{1}{2} \frac{\cosh(k+n)x}{k+n} + c.$$

WORKED EXAMPLE 59

(i) $\int 2 \sinh 2x \cosh 3x \, dx$

(ii) $\int \cosh 3x \cosh 4x \, dx.$

Solution 59

(i) $\int 2 \sinh 2x \cosh 3x \, dx$

$$= \int [\sinh(2x - 3x) + \sinh(2x + 3x)] \, dx$$

$$= \int [\sinh(-x) + \sinh 5x] \, dx$$

$$= \int (-\sinh x + \sinh 5x) \, dx$$

$$= -\cosh x + \frac{1}{5} \cosh 5x + c$$

where $\sinh(-x) = \dfrac{e^{-x} - e^{-(-x)}}{2}$

$$= \frac{e^{-x} - e^x}{2}$$

$$= -\frac{e^x - e^{-x}}{2}$$

$$= -\sinh x \text{ an odd function.}$$

(ii) $\int \cosh 3x \cosh 4x \, dx$

$\cosh(x+y) = \cosh x \cosh y + \sinh x \sinh y$

since $\cos(x+y) = \cos x \cos y - \sin x \sin y$

$\cosh(x+y) = \cosh x \cosh y$

$\qquad - i \sinh x \, i \sinh y$

$\qquad = \cosh x \cosh y + \sinh x \sinh y$

replacing $\sin x$ by $i \sinh x$ in the circular function identities. Similarly

$\cosh(x-y) = \cosh x \cosh y - \sinh x \sinh y$

$\cosh(x+y) + \cosh(x-y) = 2 \cosh x \cosh y$

$$\int \cosh 3x \cosh 4x \, dx$$

$$= \int \frac{1}{2}[\cosh(3x+4x) + \cosh(3x-4x)] \, dx$$

$$= \int \frac{1}{2}[\cosh 7x + \cosh(-x)] \, dx$$

$$= \frac{1}{2}\int (\cosh 7x + \cosh x) \, dx$$

$$= \frac{\sinh 7x}{14} + \frac{\sinh x}{2} + c$$

$$\cosh(-x) = \frac{e^{-x} + e^{-(-x)}}{2}$$

$$= \frac{e^x + e^{-x}}{2}$$

$$= \cosh x \text{ an even function.}$$

$$\int \sinh^3 x \, dx = \int \sinh x \sinh^2 x \, dx$$

$$= \int \sinh^2 x \, d(\cosh x)$$

$$= \int \left(\cosh^2 x - 1\right) d(\cosh x)$$

$$= \frac{\cosh^3 x}{3} - \cosh x + c$$

$$\int \cosh^5 x \, dx$$

$$= \int \cosh x \cosh^4 x \, dx$$

$$= \int \cosh^4 x \, d(\sinh x)$$

$$= \int \left(1 + 2\sinh^2 x + \sinh^4 x\right) d(\sinh x)$$

$$= \sinh x + \frac{2}{3}\sinh^3 x + \frac{1}{5}\sinh^5 x + c$$

$$\cosh^4 x = \left(1 + \sinh^2 x\right)^2$$

$$= 1 + 2\sinh^2 x + \sinh^4 x.$$

WORKED EXAMPLE 60

(i) $\int \dfrac{\cosh x}{\sinh^3 x} \, dx$

(ii) $\int \dfrac{\sinh x}{\sqrt{\cosh x}} \, dx.$

Solution 60

(i) $\int \dfrac{\cosh x}{\sinh^3 x} \, dx = \int \dfrac{d(\sinh x)}{\sinh^3 x}$

$$= \int (\sinh x)^{-3} d(\sinh x)$$

$$= \frac{(\sinh x)^{-2}}{-2} = -\frac{1}{2\sinh^2 x} + c.$$

$$= -\frac{1}{2}\operatorname{cosech}^2 x + c$$

(ii) $\int \dfrac{\sinh x}{\sqrt{\cosh x}} \, dx = \int \dfrac{d(\cosh x)}{\sqrt{\cosh x}}$

$$= \int (\cosh x)^{-\frac{1}{2}} d(\cosh x)$$

$$= 2(\cosh x)^{\frac{1}{2}} + c.$$

$$\int e^x \sinh 2x \, dx = \frac{\cosh 2x}{2} e^x - \int \frac{\cosh 2x}{2} e^x \, dx$$

$$= \frac{1}{2}\cosh 2x \, e^x - \left[\frac{1}{4}\sinh 2x \, e^x\right.$$

$$\left. - \int \frac{1}{4}\sinh 2x \, e^x \, dx\right]$$

$$I = \frac{1}{2}\cosh 2x \, e^x$$

$$- \frac{1}{4}\sinh 2x \, e^x + \frac{1}{4}I$$

$$\frac{3}{4}I = \frac{1}{2}\cosh 2x \, e^x - \frac{1}{4}\sinh 2x \, e^x$$

$$I = \int e^x \sinh 2x \, dx$$

Hyperbolic Functions

$$= \frac{4}{3}\left(\frac{1}{2}\cosh 2x\, e^x\right.$$

$$\left. -\frac{1}{4}\sinh 2x\, e^x\right)$$

$$= \frac{2}{3}e^x \cosh 2x$$

$$-\frac{1}{3}\sinh 2x\, e^x + c.$$

Alternatively

$$\int e^x \sinh 2x\, dx = \frac{1}{2}\int e^x \left(e^{2x} - e^{-2x}\right) dx$$

$$= \frac{1}{2}\int \left(e^{3x} - e^{-x}\right) dx$$

$$= \frac{1}{6}e^{3x} + \frac{1}{2}e^{-x} + c.$$

This is much easier than the previous method by parts. Show that

$$\frac{2}{3}e^x \cosh 2x - \frac{1}{3}e^x \sinh 2x = \frac{1}{6}e^{3x} + \frac{1}{2}e^{-x}.$$

WORKED EXAMPLE 61

Evaluate the definite integrals.

(i) $\int_1^2 e^x \sinh x\, dx$

(ii) $\int_1^2 e^x \cosh x\, dx$.

Solution 61

(i) $\int_1^2 e^x \sinh x\, dx = \frac{1}{2}\int_1^2 e^x \left(e^x - e^{-x}\right) dx$

$$= \frac{1}{2}\int_1^2 \left(e^{2x} - 1\right) dx$$

$$= \left[\frac{e^{2x}}{4} - \frac{1}{2}x\right]_1^2$$

$$= \frac{e^4}{4} - 1 - \frac{e^2}{4} + \frac{1}{2}$$

$$= \frac{e^4}{4} - \frac{e^2}{4} - \frac{1}{2}$$

$$= 13.65 - 1.85 - 0.5 = 11.3$$

(ii) $\int_1^2 e^x \cosh x\, dx = \frac{1}{2}\int_1^2 e^x \left(e^x + e^{-x}\right) dx$

$$= \frac{1}{2}\int_1^2 \left(e^{2x} + 1\right) dx$$

$$= \left[\frac{e^{2x}}{4} + \frac{1}{2}x\right]_1^2$$

$$= \frac{e^4}{4} + 1 - \frac{e^2}{4} - \frac{1}{2}$$

$$= 13.65 - 1.85 + 0.5 = 12.3.$$

WORKED EXAMPLE 62

Show that $\int_0^2 \sqrt{1 + \sinh^2 2x}\, dx = \frac{1}{2}\sinh 4.$

Solution 62

$$\int_0^2 \sqrt{1 + \sinh^2 2x}\, dx = \int_0^2 \cosh 2x\, dx$$

$$= \left[\frac{\sinh 2x}{2}\right]_0^2$$

$$= \frac{1}{2}\sinh 4.$$

where $\cosh^2 2x - \sinh^2 2x = 1$, $\cosh^2 2x = 1 + \sinh^2 2x$.

WORKED EXAMPLE 63

Show that the definite integral

$$\int_0^1 \frac{\sinh^2 x}{\cosh x}\, dx \text{ is equal to } 1.175 - 2\tan^{-1} e + \frac{\pi}{2}.$$

Solution 63

$$\int_0^1 \frac{\sinh^2 x}{\cosh x} \, dx = \int_0^1 \frac{\cosh^2 x - 1}{\cosh x} \, dx$$

$$= \int_0^1 (\cosh x - \operatorname{sech} x) \, dx$$

$$= \int_0^1 \cosh x \, dx - \int_0^1 \operatorname{sech} x \, dx$$

$$= [\sinh x]_0^1 - \int_0^1 \frac{2}{e^x + e^{-x}} \, dx$$

$$= 1.175 - \int_0^1 \frac{2}{e^x + e^{-x}} \, dx$$

$$\int_0^1 \frac{2}{e^x + e^{-x}} \, dx \text{ let } e^x = u, \, \frac{du}{dx} = e^x, \, dx = \frac{du}{e^x} = \frac{du}{u}$$

$$\int_1^e \frac{2}{\left(u + \frac{1}{u}\right)} \frac{du}{u} = \int_1^e \frac{2}{u^2 + 1} \, du = \left[2 \tan^{-1} u\right]_1^e$$

$$= 2 \tan^{-1} e - 2 \tan^{-1} 1$$

$$= 2 \tan^{-1} e - \frac{2\pi}{4}$$

$$= 2 \tan^{-1} e - \frac{\pi}{2}$$

$$\therefore \int_0^1 \frac{\sinh^2 x}{\cosh x} \, dx = 1.175 - 2 \tan^{-1} e + \frac{\pi}{2}.$$

WORKED EXAMPLE 64

Evaluate the definite integrals

(i) $\int_0^1 \sqrt{1 + 9x^2} \, dx$

(ii) $\int_0^1 \cosh^2 x \, dx$

(iii) $\int_{0.5}^1 \operatorname{cosech} x \, dx$

(iv) $\int_0^1 x \sinh x \, dx$

(v) $\int_0^1 \frac{dx}{\sqrt{(x+1)^2 + 1}}.$

Solution 64

(i) $\int_0^1 \sqrt{1 + 9x^2} \, dx$

$$= \int_0^{\sinh^{-1} 3} \sqrt{1 + \sinh^2 y} \cdot \frac{1}{3} \cosh y \, dy$$

let $\sinh y = 3x$

$$\cosh y \frac{dy}{dx} = 3$$

$$= \frac{1}{3} \int_0^{\sinh^{-1} 3} \cosh^2 y \, dy$$

$$= \frac{1}{3} \int_0^{\sinh^{-1} 3} \frac{(\cosh 2y + 1)}{2} \, dy$$

$$= \frac{1}{6} \left[\frac{\sinh 2y}{2} + y\right]_0^{\sinh^{-1} 3}$$

where $\cosh 2y = 2 \cosh^2 y - 1$,

$$\frac{\cosh 2y + 1}{2} = \cosh^2 y$$

$$= \frac{1}{6} \left[\frac{\sinh 2 \sinh^{-1} 3}{2} + \sinh^{-1} 3\right]$$

$$= \frac{1}{6} (9.486833 + 1.8184465) = 1.88.$$

(ii) $\int_0^1 \cosh^2 x \, dx = \frac{1}{2} \int_0^1 (\cosh 2x + 1) \, dx$

$$= \left[\frac{\sinh 2x}{4} + \frac{x}{2}\right]_0^1 = \frac{\sinh 2}{4} + \frac{1}{2} = 1.41$$

(iii) $\displaystyle\int_{0.5}^{1} \operatorname{cosech} x \, dx = \int_{0.5}^{1} \frac{1}{\sinh x} \, dx$

$\displaystyle = 2\int_{0.5}^{1} \frac{1}{e^x - e^{-x}} \, dx$

let $u = e^x$, $\dfrac{du}{dx} = e^x$, $dx = \dfrac{du}{e^x}$

$\displaystyle 2\int_{0.5}^{1} \frac{1}{e^x - e^{-x}} \, dx = 2\int_{0.5}^{e} \frac{du}{(u^2 - 1)}$

$\displaystyle = \int_{0.5}^{e} \left(\frac{1}{u-1} - \frac{1}{u+1} \right) du$

$\dfrac{1}{u^2 - 1} = \dfrac{1}{(u-1)} \cdot \dfrac{1}{(u+1)}$

$= \dfrac{A}{u-1} + \dfrac{B}{u+1} = \dfrac{\frac{1}{2}}{u-1} - \dfrac{\frac{1}{2}}{u+1}$

$A(u+1) + B(u-1) \equiv 1$

$u = 1, A = \dfrac{1}{2} \qquad u = -1, B = -\dfrac{1}{2}$

$\displaystyle \int_{e^{0.5}}^{e} \left(\frac{1}{u-1} - \frac{1}{u+1} \right) du$

$= \left[\ln(u-1) - \ln(u+1) \right]_{e^{0.5}}^{e}$

$= \ln \dfrac{e-1}{e+1} - \ln \dfrac{e^{0.5}-1}{e^{0.5}+1}$

$= \ln 0.462 - \ln 0.245 = 0.636.$

(iv) $\displaystyle \int_0^1 x \sinh x \, dx = [x \cosh x]_0^1 - \int_0^1 \cosh x \, dx$

$= \cosh 1 - [\sinh x]_0^1$

$= \cosh 1 - \sinh 1$

$= 1.543 - 1.175 = 0.368.$

(v) $\displaystyle \int_0^1 \frac{dx}{\sqrt{(x+1)^2 + 1}} \qquad \sinh^2 y + 1 = \cosh^2 y$

let $x + 1 = \sinh y$, $\dfrac{dx}{dy} = \cosh y$

$\displaystyle \int_{\sinh^{-1} 1}^{\sinh^{-1} 2} \frac{\cosh y \, dy}{\sqrt{\sinh^2 y + 1}} = [y]_{\sinh^{-1} 1}^{\sinh^{-1} 2}$

$= \sinh^{-1} 2 - \sinh^{-1} 1$

$= 1.444 - 0.881 = 0.563.$

WORKED EXAMPLE 65

Determine the following integrals by making a hyperbolic substitution.

(i) $\displaystyle \int \sqrt{(x+2)^2 + 4} \, dx$

(ii) $\displaystyle \int \sqrt{x^2 + 2x + 10} \, dx$

(iii) $\displaystyle \int \sqrt{4x^2 + 25} \, dx$

(iv) $\displaystyle \int \frac{1}{\sqrt{x^2 + 5}} \, dx$

(v) $\displaystyle \int \frac{1}{\sqrt{x^2 - 2x + 17}} \, dx.$

Solution 65

(i) $\displaystyle \int \sqrt{(x+2)^2 + 4} \, dx$

let $x + 2 = 2 \sinh y$, $\dfrac{dx}{dy} = 2 \cosh y$

$\displaystyle \int \sqrt{(4 \sinh^2 y) + 4} \, dx$

$\displaystyle = 2 \int \sqrt{\sinh^2 y + 1} \, 2 \cosh y \, dy$

$\displaystyle = 4 \int \cosh^2 y \, dy$

$$\frac{\cosh 2y + 1}{2} = \cosh^2 y$$

$$= 2\int (\cosh 2y + 1)\, dy$$

$$= 2\left(\frac{\sinh 2y}{2} + y\right) + c$$

$$= \sinh 2y + 2y + c$$

$$= 2\sinh y \cosh y + 2y + c$$

$$= 2\frac{(x+2)}{2}\sqrt{\left(\frac{x+2}{2}\right)^2 + 1}$$

$$+ 2\sinh^{-1}\frac{x+2}{2} + c$$

$$= \frac{1}{2}(x+2)\sqrt{x^2 + 4x + 8}$$

$$+ 2\sinh^{-1}\frac{x+2}{2} + c.$$

(ii) $\int \sqrt{x^2 + 2x + 10}\, dx = \int \sqrt{(x+1)^2 + 3^2}\, dx$

let $x + 1 = 3\sinh y$, $\dfrac{dx}{dy} = 3\cosh y$

$$\int \sqrt{9\sinh^2 y + 9}\, dx = 3\int \cosh y \cdot 3\cosh y\, dy$$

$$= 9\int \cosh^2 y\, dy = \frac{9}{2}\int (\cosh 2y + 1)\, dy$$

$$= \frac{9}{2}\left(\frac{\sinh 2y}{2} + y\right) = \frac{9}{2}(\sinh y \cosh y + y)$$

$$= \frac{9}{2}\left(\frac{x+1}{3}\sqrt{\left(\frac{x+1}{3}\right)^2 + 1}\right.$$

$$\left. + \sinh^{-1}\left(\frac{x+1}{3}\right)\right) + c$$

$$= \frac{1}{2}(x+1)\sqrt{x^2 + 2x + 10}$$

$$+ \frac{9}{2}\sinh^{-1}\frac{(x+1)}{3} + c.$$

(iii) $\int \sqrt{4x^2 + 25}\, dx \quad \sinh^2 y + 1 = \cosh^2 y$

$$= \int \sqrt{(2x)^2 + 5^2}\, dx$$

let $2x = 5\sinh y$

$$= \int \sqrt{25\sinh^2 y + 25} \cdot \frac{5}{2}\cosh y\, dy$$

$$\frac{dx}{dy} = \frac{5}{2}\cosh y$$

$$= \frac{25}{2}\int \cosh^2 y\, dy = \frac{25}{4}\int (\cos 2y + 1)\, dy$$

$$= \frac{25}{4}\frac{\sinh 2y}{2} + \frac{25}{4}y + c$$

$$= \frac{25}{4}\sinh y \cosh y + \frac{25}{4}y + c$$

$$= \frac{25}{4}\frac{2x}{5}\sqrt{1 + \left(\frac{2x}{5}\right)^2} + \frac{25}{4}\sinh^{-1}\frac{2x}{5} + c$$

$$= \frac{1}{2}x\sqrt{25 + 4x^2} + \frac{25}{4}\sinh^{-1}\frac{2x}{5} + c.$$

(iv) $\int \dfrac{1}{\sqrt{x^2 + 5}}\, dx \quad$ let $x = \sqrt{5}\sinh y$

$$\frac{dx}{dy} = \sqrt{5}\cosh y$$

$$\int \frac{1}{\sqrt{5}\sqrt{\sinh^2 y + 1}}\sqrt{5}\cosh y\, dy = y$$

$$= \sinh^{-1}\frac{x}{\sqrt{5}}.$$

(v) $\int \dfrac{1}{\sqrt{x^2 - 2x + 17}}\, dx = \int \dfrac{1}{\sqrt{(x-1)^2 + 4^2}}\, dx$

let $x - 1 = 4\sinh y$, $\dfrac{dx}{dy} = 4\cosh y$

$$\int \frac{1}{4\sqrt{\sinh^2 y + 1}}4\cosh y\, dy = y = \sinh^{-1}\frac{x-1}{4}$$

$$\int \frac{1}{\sqrt{x^2 - 2x + 17}}\, dx = \sinh^{-1}\frac{x-1}{4} + c.$$

Exercises 6

1. $\int_0^2 \sinh 3x \, dx$

2. $\int_1^4 \cosh 4x \, dx$

3. $\int_2^3 \tanh 5x \, dx$

4. $\int_1^2 \coth 2x \, dx$

5. $\int_{\frac{1}{2}}^{\frac{3}{4}} \operatorname{cosech} \frac{1}{2}x \, dx$

6. $\int_0^1 \operatorname{sech} \frac{1}{3}x \, dx$.

7. (a)

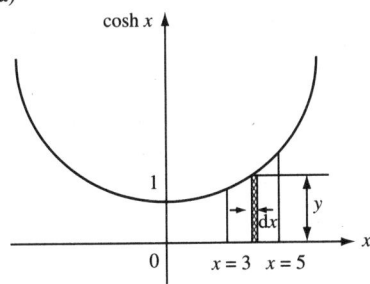

Fig. 7-I/25

(b)

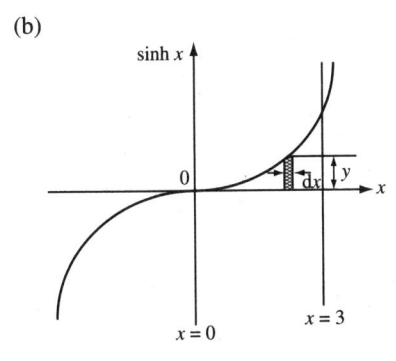

Fig. 7-I/26

Determine the areas under the curves between the limits indicated.

8. (i) $\int_0^1 \tanh 3x \, dx$

   (ii) $\int_1^2 \operatorname{sech} 2x \, dx$

   (iii) $\int_1^2 \operatorname{cosech} \frac{3}{4}x \, dx$.

9. Write down the integrals with respect to each variable

   (i) $\dfrac{1}{\sqrt{x^2 - 9}}$

   (ii) $\dfrac{1}{\sqrt{x^2 + 25}}$

   (iii) $\dfrac{1}{\sqrt{9x^2 + 25}}$

   (iv) $\dfrac{1}{\sqrt{25x^2 - 9}}$

   (v) $\dfrac{1}{\sqrt{4x^2 - 25}}$.

10. Prove, by using a suitable substitution the following:

    (i) $\int \dfrac{1}{\sqrt{x^2 - a^2}} \, dx = \operatorname{arcosh} \dfrac{x}{a} + c$

    (ii) $\int \dfrac{1}{\sqrt{a^2 x^2 + b^2}} \, dx = \dfrac{1}{a} \operatorname{arcsinh} \dfrac{x}{\frac{b}{a}} + c$.

11. Integrate the following with respect to x:

    (i) $\int \operatorname{sech}^2 2x \, dx$

    (ii) $\int \operatorname{cosech}^2 3x \, dx$

    (iii) $\int \tanh^2 \dfrac{1}{2}x \, dx$

    (iv) $\int \coth^2 \dfrac{3}{2}x \, dx$

    (v) $\int \cosh^2 5x \, dx$

    (vi) $\int \sinh^2 7x \, dx$.

12. Determine the integrals:

   (i) $\int \sinh 7x \cosh 5x \, dx$

   (ii) $\int \sinh 3x \cosh 7x \, dx$

   (iii) $\int \cosh 2x \cosh 3x \, dx$

   (iv) $\int \cosh 9x \cosh 5x \, dx.$

13. Determine the integrals:

   (i) $\int \dfrac{\sinh x}{\sqrt[4]{\cosh^3 x}} \, dx$

   (ii) $\int \dfrac{\cosh x}{\sinh^5 x} \, dx$

Determine the integrals:

14. $\int \dfrac{dx}{\sqrt{1 + 16x^2}}$

15. $\int \sinh^2 4x \, dx$

16. $\int e^{2x} \cosh x \, dx$

17. $\int e^x \cosh x \, dx$

18. $\int e^{-x} \sinh x \, dx$

19. $\int e^x \sinh x \, dx$

20. $\int e^{3x} \cosh 2x \, dx$

21. $\int e^{2x} \sinh 2x \, dx.$

22. Evaluate the integrals:

   (i) $\int_0^1 e^x \sinh x \, dx$

   (ii) $\int_0^1 e^x \cosh x \, dx.$

23. Find the integrals:

   (i) $\int \dfrac{dx}{\sqrt{x^2 - 1}}$

   (ii) $\int \dfrac{dx}{\sqrt{x^2 + 1}}$

   (iii) $\int \dfrac{dx}{\sqrt{x^2 - a^2}}$

   (iv) $\int \dfrac{dx}{\sqrt{x^2 + a^2}}.$

24. Find the integrals:

   (i) $\int \dfrac{dx}{3 + 4\cosh x}$

   (ii) $\int \dfrac{dx}{\sinh x - \cosh x}$

   (iii) $\int \dfrac{\sinh x \, dx}{2 \sinh x - \cosh x}.$

25. Express the following:

   (i) $2 \sinh x + \cosh x$

   (ii) $5 \sinh x + 3 \cosh x$

   (iii) $13 \cosh x + 5 \sinh x$

   in the form $R \sinh (x \pm \alpha)$ or $R \cosh (x \pm \alpha)$ giving α in logarithmic form.

26. Evaluate the following integrals in terms of e.

   (i) $\int_0^2 \dfrac{dx}{2 \sinh x + \cosh x}$

   (ii) $\int_0^2 \dfrac{dx}{5 \sinh x + 3 \cosh x}$

   (iii) $\int_0^1 \dfrac{dx}{13 \cosh x + 5 \sinh x}.$

27. Using the definitions
$$\cosh x = \frac{e^x + e^{-x}}{2}$$
$$\sinh x = \frac{e^x - e^{-x}}{2}$$
show the following fundamental identities:
   (i) $\cosh^2 x - \sinh^2 x = 1$
   (ii) $1 - \tanh^2 x = \text{sech}^2 x$
   (iii) $1 - \coth^2 x = -\text{cosech}^2 x$
   (iv) $\sinh(x + y)$
   $\quad = \sinh x \cosh y + \sinh y \cosh x$
   (v) $\cosh(x - y)$
   $\quad = \cosh x \cosh y - \sinh x \sinh y$
   (vi) $\tanh 2x = \dfrac{2 \tanh x}{1 + \tanh^2 x}$
   (vii) $\sinh 2x = 2 \sinh x \cosh x$
   (viii) $\cosh 2x = 2 \cosh^2 x - 1$
   (ix) $\cosh 4x = 1 + 2 \sinh^2 2x$
   (x) $\sinh 2x = \dfrac{2 \tanh x}{1 - \tanh^2 x}$.

7

Integration by Parts

Let u and v be two functions of x.

$$y = uv$$

$$\frac{dy}{dx} = \frac{du}{dx}v + u\frac{dv}{dx}$$

$$u\frac{dv}{dx} = \frac{d}{dx}(uv) - v\frac{du}{dx}.$$

Integrating each term with respect to x

$$\int u\frac{dv}{dx}dx = \int \frac{d}{dx}(uv)dx - \int v\frac{du}{dx}dx$$

$$\int \underset{2}{u}\,\underset{1}{dv} = uv - \int v\,du \qquad \ldots(1)$$

Observe equation (1) which is the formula for integrating by parts. The integral of $u\,dv$, the product of two functions of x which do not bear any relationship is equal to uv, dv is integrated, minus the integral $v\,du$ where v is the integrated function of v times du, the derivative of u. The integral of dv is denoted by **1** and the derivative of u is denoted by **2**.

Formula (1) may be written as

$$\boxed{\int u\,dv = uv - \int v\,du}$$

1 denotes that dv is first integrated and **2** denotes that u is secondly differentiated. Let us try to apply this method.

WORKED EXAMPLE 66

(i) $\int x\,e^x\,dx$

(ii) $\int x\sin x\,dx$

(iii) $\int e^x \sin x\,dx$

Solution 66

(i) $\int x\,e^x\,dx$. The product of the two functions x and e^x is shown, x is an algebraic functions and e^x is an exponential function, these two functions has no direct relationship, so by parts method is the most evident method. We can mark e^x by **1** and x by **2**.

$$\int \underset{2}{x}\,\underset{1}{e^x}\,dx = e^x \cdot x - \int e^x \cdot 1\,dx = x\,e^x - e^x + c.$$

What happens if we denote x by **1** and e^x by **2**?

$$\int \underset{1}{x}\,\underset{2}{e^x}\,dx = \frac{x^2}{2}e^x - \int \frac{x^2}{2}e^x\,dx,$$ it can be seen that this way, the indefinite integral $\int \frac{x^2}{2}e^x\,dx$ is now more difficult.

(ii) Try the next example

$$\int \underset{2}{x}\,\underset{1}{\sin x}\,dx = (-\cos x)x - \int (-\cos x)\cdot 1 \cdot dx$$

$$= -x\cos x + \sin x + c.$$

Again if we denote the functions the other way the integral will not be found.

(iii) $\int \underset{2}{e^x}\,\underset{1}{\sin x}\,dx = -\cos x\,e^x - \int -\cos x\,e^x\,dx$

$$= -e^x \cos x + \int \underset{1}{\cos x}\,\underset{2}{e^x}\,dx$$

$$\int e^x \sin x\,dx = -e^x \cos x + \sin x\,e^x$$

$$- \int \sin x\,e^x\,dx$$

Let $I = \int e^x \sin x\,dx$

$$I = -e^x \cos x + e^x \sin x - I$$

$2I = -e^x \cos x + e^x \sin x$

$I = -\dfrac{1}{2}e^x \cos x + \dfrac{1}{2}e^x \sin x + c$

$\displaystyle\int e^x \sin x \, dx = -\dfrac{1}{2}e^x \cos x + \dfrac{1}{2}e^x \sin x + c$

this is a little more difficult, it resulted in applying the by parts formula twice, but again, care must be taken to denote in each case the trigonometric function by **1**.

WORKED EXAMPLE 67

Integrate the following indefinite integrals:

(i) $\displaystyle\int e^x \cos x \, dx$

(ii) $\displaystyle\int e^{-x} \cos 2x \, dx$

(iii) $\displaystyle\int e^{3x} \sin 3x \, dx.$

Solution 67

(i) $\displaystyle\int \underset{2}{e^x} \underset{1}{\cos x} \, dx = \sin x \, e^x - \int \sin x \, e^x \, dx$

$= \sin x \, e^x - \left[-\cos x \, e^x - \int -\cos x \, e^x \, dx \right]$

$\displaystyle\int e^x \cos x \, dx = e^x \sin x + e^x \cos x - \int e^x \cos x \, dx$

let $\displaystyle\int e^x \cos x \, dx = I$

$I = e^x \sin x + e^x \cos x - I$

$2I = e^x \sin x + e^x \cos x$

$\boxed{I = \displaystyle\int e^x \cos x \, dx = \dfrac{1}{2}e^x \sin x + \dfrac{1}{2}e^x \cos x + c}$

(ii) $\displaystyle\int \underset{2}{e^{-x}} \underset{1}{\cos 2x} \, dx$

$= \dfrac{\sin 2x}{2} e^{-x} - \int \dfrac{\sin 2x}{2} (-e^{-x}) \, dx$

$= \dfrac{1}{2} \sin 2x \, e^{-x} + \dfrac{1}{2} \int e^{-x} \sin 2x \, dx$

$= \dfrac{1}{2} \sin 2x \, e^{-x} + \dfrac{1}{2} \left[\dfrac{-\cos 2x}{2} e^{-x} \right.$

$\left. - \int \dfrac{-\cos 2x}{2} (-e^{-x}) \, dx \right]$

$= \dfrac{1}{2} e^{-x} \sin 2x - \dfrac{1}{4} e^{-x} \cos 2x - \dfrac{1}{4} \int e^{-x} \cos 2x \, dx$

$\dfrac{5}{4} \displaystyle\int e^{-x} \cos 2x \, dx = \dfrac{1}{2} e^{-x} \sin 2x - \dfrac{1}{4} e^{-x} \cos 2x$

$\boxed{\displaystyle\int e^{-x} \cos 2x \, dx = \dfrac{2}{5} e^{-x} \sin 2x - \dfrac{1}{5} e^{-x} \cos 2x + c}$

(iii) $\displaystyle\int \underset{2}{e^{3x}} \underset{1}{\sin 3x} \, dx$

$= \dfrac{-\cos 3x}{3} e^{3x} - \int \dfrac{-\cos 3x}{3} 3 e^{3x} \, dx$

$= -\dfrac{1}{3} e^{3x} \cos 3x + \int \underset{2}{e^{3x}} \underset{1}{\cos 3x} \, dx$

$= -\dfrac{1}{3} e^{3x} \cos 3x + \dfrac{\sin 3x}{3} e^{3x} - \int \dfrac{\sin 3x}{3} 3 e^{3x} \, dx$

$= -\dfrac{1}{3} e^{3x} \cos 3x + \dfrac{\sin 3x}{3} e^{3x} - \int e^{3x} \sin 3x \, dx$

$\boxed{\displaystyle\int e^{3x} \sin 3x \, dx = -\dfrac{1}{6} e^{3x} \cos 3x + \dfrac{1}{6} e^{3x} \sin 3x + c}$

Integrate (i) $\displaystyle\int e^{ax} \sin bx \, dx$ (ii) $\displaystyle\int e^{ax} \cos bx \, dx$

(i) $\displaystyle\int \underset{2}{e^{ax}} \underset{1}{\sin bx} \, dx$

$= \dfrac{-\cos bx}{b} e^{ax} - \int \dfrac{-\cos bx}{b} a e^{ax} \, dx$

$= -\dfrac{1}{b} e^{ax} \cos bx + \dfrac{a}{b} \int \underset{2}{e^{ax}} \underset{1}{\cos bx} \, dx$

$$= -\frac{1}{b}e^{ax}\cos bx + \frac{a}{b}\left[\frac{\sin bx}{b}e^{ax}\right.$$

$$\left. - \int \frac{\sin bx}{b}ae^{ax}\,dx\right]$$

$$= -\frac{1}{b}e^{ax}\cos bx + \frac{a}{b^2}e^{ax}\sin bx$$

$$- \frac{a^2}{b^2}\int e^{ax}\sin bx\,dx$$

$$\left(1+\frac{a^2}{b^2}\right)\int e^{ax}\sin bx\,dx = -\frac{1}{b}e^{ax}\cos bx$$

$$+ \frac{a}{b^2}e^{ax}\sin bx$$

$$\boxed{\int e^{ax}\sin bx\,dx = \frac{-b}{a^2+b^2}e^{ax}\cos bx + \frac{a}{a^2+b^2}e^{ax}\sin bx}$$

(ii) $\int \underset{2}{e^{ax}}\underset{1}{\cos bx}\,dx = \frac{\sin bx}{b}e^{ax} - \int \frac{\sin bx}{b}ae^{ax}\,dx$

$$= \frac{1}{b}e^{ax}\sin bx - \frac{a}{b}\int \underset{2}{e^{ax}}\underset{1}{\sin bx}\,dx$$

$$= \frac{1}{b}e^{ax}\sin bx - \frac{a}{b}\left[\frac{-\cos bx}{b}e^{ax}\right.$$

$$\left. - \int \frac{-\cos bx}{b}ae^{ax}\,dx\right]$$

$$= \frac{1}{b}e^{ax}\sin bx + \frac{a}{b^2}e^{ax}\cos bx$$

$$- \frac{a^2}{b^2}\int e^{ax}\cos bx\,dx$$

$$\left(1+\frac{a^2}{b^2}\right)\int e^{ax}\cos bx\,dx = \frac{1}{b}e^{ax}\sin bx$$

$$+ \frac{a}{b^2}e^{ax}\cos bx$$

$$\boxed{\int e^{ax}\cos bx\,dx = \frac{b}{a^2+b^2}e^{ax}\sin bx + \frac{a}{a^2+b^2}e^{ax}\cos bx + c}$$

Integrate $\int \log_e x\,dx$

$$\int \underset{2}{\ln x}\,\underset{1}{dx} = x\ln x - \int x\cdot\frac{1}{x}\,dx = x\ln x - \int dx$$

$$\boxed{\int \ln x\,dx = x\ln x - x + c}$$

WORKED EXAMPLE 68

Integrate by parts the following:

(i) $\int x\ln 7x\,dx$

(ii) $\int x^2 \ln 5x\,dx$

(iii) $\int x^3 \ln 4x\,dx$.

Solution 68

(i) $\int \underset{1}{x}\underset{2}{\ln 7x}\,dx = \frac{x^2}{2}\ln 7x - \int \frac{x^2}{2}\cdot\frac{1}{x}\,dx$

$$= \frac{x^2}{2}\ln 7x - \frac{1}{2}\int x\,dx$$

$$= \frac{x^2}{2}\ln 7x - \frac{1}{2}\frac{x^2}{2} + c$$

$$\boxed{\int x\ln 7x\,dx = \frac{1}{2}x^2 \ln 7x - \frac{1}{4}x^2 + c}$$

(ii) $\int \underset{1}{x^2}\underset{2}{\ln 5x}\,dx = \frac{x^3}{3}\ln 5x - \int \frac{x^3}{3}\cdot\frac{1}{x}\,dx$

$$= \frac{1}{3}x^3 \ln 5x - \frac{1}{3}\int x^2\,dx$$

$$\boxed{\int x^2 \ln 5x\,dx = \frac{1}{3}x^3 \ln 5x - \frac{1}{9}x^3 + c}$$

(iii) $\int x^3 \ln 4x\,dx = \frac{x^4}{4}\ln 4x - \int \frac{x^4}{4}\cdot\frac{1}{x}\,dx$

$$= \frac{1}{4}x^4 \ln 4x - \frac{1}{4}\int x^3\,dx$$

$$\boxed{\int x^3 \ln 4x\,dx = \frac{1}{4}x^4 \ln 4x - \frac{1}{16}x^4 + c}$$

Integrate

(i) $\int \sin^{-1} x \, dx$

(ii) $\int \cos^{-1} x \, dx$

(iii) $\int \tan^{-1} x \, dx$

(iv) $\int \cot^{-1} x \, dx$

(v) $\int \sec^{-1} x \, dx$

(vi) $\int \text{cosec}^{-1} x \, dx$.

(i) $\int \underset{2}{\sin^{-1}} \underset{1}{x} \, dx$

$= x \sin^{-1} x - \int x \cdot \dfrac{1}{(1-x^2)^{\frac{1}{2}}} \, dx$

$= x \sin^{-1} x + \dfrac{1}{2} \int \dfrac{d(1-x^2)}{(1-x^2)^{\frac{1}{2}}} \, dx$

$= x \sin^{-1} x + \dfrac{1}{2} \int (1-x^2)^{-\frac{1}{2}} d(1-x^2)$

$\boxed{\int \sin^{-1} \, dx = x \sin^{-1} x + (1-x^2)^{\frac{1}{2}} + c}$

(ii) $\int \underset{2}{\cos^{-1}} \underset{1}{x} \, dx$

$= x \cos^{-1} x - \int x \cdot \left(-\dfrac{1}{(1-x^2)^{\frac{1}{2}}}\right) dx$

$= x \cos^{-1} x - \int \dfrac{1}{2} \dfrac{d(1-x^2)}{(1-x^2)^{\frac{1}{2}}}$

$\boxed{\int \cos^{-1} x \, dx = x \cos^{-1} x - (1-x^2)^{\frac{1}{2}} + c}$

(iii) $\int \underset{2}{\tan^{-1}} \underset{1}{x} \, dx = x \tan^{-1} x - \int x \cdot \dfrac{1}{1+x^2} \, dx$

$= x \tan^{-1} x - \dfrac{1}{2} \int \dfrac{d(1+x^2)}{1+x^2}$

$\boxed{\int \tan^{-1} x \, dx = x \tan^{-1} x - \dfrac{1}{2} \ln(1+x^2) + c}$

(iv) $\int \underset{2}{\cot^{-1}} \underset{1}{x} \, dx = x \cot^{-1} x - \int x \left[-\dfrac{1}{1+x^2}\right] dx$

$= x \cot^{-1} x + \dfrac{1}{2} \int \dfrac{d(1+x^2)}{1+x^2}$

$\boxed{\int \cot^{-1} x \, dx = x \cot^{-1} x + \dfrac{1}{2} \ln(1+x^2) + c}$

(v) $\int \underset{2}{\sec^{-1}} \underset{1}{x} \, dx$

$= x \sec^{-1} x - \int x \cdot \dfrac{1}{x(x^2-1)^{\frac{1}{2}}} dx$

$= x \sec^{-1} x - \int \dfrac{1}{(x^2-1)^{\frac{1}{2}}} dx$

$\boxed{\int \sec^{-1} x \, dx = x \sec^{-1} x - \cosh^{-1} x + c}$

(vi) $\int \underset{2}{\text{cosec}^{-1}} \underset{1}{x} \, dx$

$= x \, \text{cosec}^{-1} x - \int x \cdot \left[-\dfrac{1}{x(x^2-1)^{\frac{1}{2}}}\right] dx$

$= x \, \text{cosec}^{-1} x + \int \dfrac{1}{(x^2-1)^{\frac{1}{2}}} dx$

$\boxed{\int \text{cosec}^{-1} x \, dx = x \, \text{cosec}^{-1} x + \cosh^{-1} x + c}$.

WORKED EXAMPLE 69

Integrate by parts the following:

(i) $\sinh^{-1} x$

(ii) $\cosh^{-1} x$

(iii) $\tanh^{-1} x$

(iv) $\coth^{-1} x$

(v) $\text{sech}^{-1} x$

(vi) $\text{cosech}^{-1} x$.

Solution 69

(i) $\int \sinh^{-1} x \, dx = x \sinh^{-1} x - \int x \cdot \frac{1}{(1+x^2)} \, dx$

$= x \sinh^{-1} x - \frac{1}{2} \int \frac{d(1+x^2)}{(1+x^2)^{\frac{1}{2}}}$

$= x \sinh^{-1} x - (1+x^2)^{\frac{1}{2}} + c.$

(ii) $\int \cosh^{-1} x \, dx = x \cosh^{-1} x - \int x \cdot \frac{1}{\sqrt{x^2-1}} \, dx$

$= x \cosh^{-1} x - \frac{1}{2} \int \frac{d(x^2-1)}{(x^2-1)^{\frac{1}{2}}}$

$= x \cosh^{-1} x - (x^2-1)^{\frac{1}{2}} + c.$

(iii) $\int \tanh^{-1} x \, dx$

$= x \tanh^{-1} x - \int x \cdot \frac{1}{1-x^2} \, dx$

$= x \tanh^{-1} x + \frac{1}{2} \int \frac{d(1-x^2)}{(1-x^2)}$

$= x \tanh^{-1} x + \frac{1}{2} \ln|(1-x^2)| + c.$

(iv) $\int \coth^{-1} x \, dx$

$= x \coth^{-1} x - \int x \cdot \frac{1}{1-x^2} \, dx$

$= x \coth^{-1} x + \frac{1}{2} \int \frac{d(1-x^2)}{1-x^2}$

$= x \coth^{-1} x + \frac{1}{2} \ln|(1-x^2)| + c.$

(v) $\int \operatorname{sech}^{-1} x \, dx$

$= x \operatorname{sech}^{-1} x - \int x \left[-\frac{1}{x(1-x^2)^{\frac{1}{2}}} \right] dx$

$= x \operatorname{sech}^{-1} x + \int \frac{1}{(1-x^2)^{\frac{1}{2}}} \, dx$

$= x \operatorname{sech}^{-1} x + \sin^{-1} x + c.$

(vi) $\int \operatorname{cosech}^{-1} x \, dx$

$= x \operatorname{cosech}^{-1} x - \int x \cdot \left[-\frac{1}{x(1+x^2)^{\frac{1}{2}}} \right] dx$

$= x \operatorname{cosech}^{-1} x + \int \frac{1}{(1+x^2)^{\frac{1}{2}}} \, dx$

$= x \operatorname{cosech}^{-1} x + \sinh^{-1} x + c.$

Exercises 7

1. $\int x \sin x \, dx$

2. $\int -x^2 \sin x \, dx$

3. $\int e^x \sin x \, dx$

4. $\int x e^x \, dx$

5. $\int 2x e^{-x} \, dx$

6. $\int e^{-x} \sin x \, dx$

7. $\int x \cos x \, dx$

8. $\int x^2 \cos x \, dx$

9. $\int e^{-x} \cos \left(x - \frac{\pi}{6} \right) dx$

10. $\int \ln x \, dx$

11. $\int x^2 \sin \left(x + \frac{\pi}{4} \right) dx$

12. $\int e^x \sin 3x \, dx$

13. $\int \ln 3x \, dx$

14. $\int x \cos \left(x - \frac{\pi}{4} \right) dx$

15. $\int 2 \sin x \cos x \, e^x \, dx$

16. $\int e^{px} \sin qx \, dx$

17. $\int x^3 \ln 3x \, dx$

18. $\int \sin 2x \, e^x \, dx$

19. $\int e^{-qx} \cos px \, dx$

20. $\int x^2 \ln 2x \, dx$

21. $\int \sin x \, e^x \cos x \, dx$

22. $\int e^{-x}(1+x)^3 \, dx$

23. $\int x^2 e^{3x+1} \, dx$

24. $\int x \sin[(n+1)x] \, dx$

Integrate by parts

25. $\int \sin^2 x \, dx$

26. $\int \cos^2 x \, dx$

27. $\int \sin^3 x \, dx$

28. $\int \cos^3 x \, dx$

29. $\int x \sin \frac{x}{3} \, dx$

30. $\int \frac{1}{4} x \, e^{\frac{1}{4}x} \, dx$

31. $\int y \sec^2 y \, dy$

32. $\int 3^x x \, dx$

33. $\int \theta \sin^2 \frac{\theta}{2} \, d\theta$

34. $\int x^3 \cos x \, dx$

35. $\int x \cot^2 3x \, dx$

36. $\int \sec^3 \theta \, d\theta$

37. $\int e^{\frac{x}{2}} \sin 2x \, dx$

38. $\int (x+1) \ln \sqrt{x+1} \, dx$

39. $\int e^x \cos 3x \, dx$

40. $\int x \, 3^x \, dx$

41. $\int x \, 2^x \, dx$

42. $\int x \, 10^x \, dx.$

8

Reduction Formulae

Let $I_n = \int \sin^n x \, dx$ where I denotes integral and I_n denotes the integral to the power n, where n is a positive integer ($n \in Z^+$), $I_{n-1} = \int \sin^{n-1} x \, dx$,

$$I_{n-2} = \int \sin^{n-2} x \, dx, \quad I_0 = \int \sin^0 x \, dx = \int dx,$$

$$I_1 = \int \sin x \, dx.$$

Reduction formulae are formulae used to express an integral I_n in terms of integrals I_{n-1} or I_{n-2} which are similar integral using $n-1$ or $n-2$, reduced powers.

Let us found such a reduction formula for $\int \sin^n x \, dx$.

$$I_n = \int \sin^n x \, dx = \int \sin x \sin^{n-1} x \, dx$$

the technique here is to use integration by parts, splitting $\sin^n x$ to $\sin x \sin^{n-1} x$, we have to choose which of these has to be integrated first, obviously is the $\sin x$.

$$I_n = \int \underset{1}{\sin x} \underset{2}{\sin^{n-1} x} \, dx = (-\cos x) \sin^{n-1} x$$

$$- \int (-\cos x)(n-1) \sin^{n-2} x \cdot \cos x \, dx$$

$$= -\cos x \sin^{n-1} x + (n-1) \int \cos^2 x \sin^{n-2} x \, dx$$

$$= -\cos x \sin^{n-1} x$$

$$+ (n-1) \int \left(1 - \sin^2 x\right) \sin^{n-2} x \, dx$$

$$= -\cos x \sin^{n-1} x$$

$$+ (n-1) \int \sin^{n-2} x \, dx - (n-1) \int \sin^n x \, dx$$

$$I_n = -\cos x \sin^{n-1} x + (n-1) I_{n-2} - (n-1) I_n$$

$$I_n + (n-1) I_n = -\cos x \sin^{n-1} x + (n-1) I_{n-2}$$

$$I_n (1 + n - 1) = -\cos x \sin^{n-1} x + (n-1) I_{n-2}$$

$$I_n = -\frac{1}{n} \cos x \sin^{n-1} x + \frac{n-1}{n} I_{n-2}.$$

Note that the arbitrary constant is omitted for convenience

$$\boxed{\begin{aligned} I_n &= \int \sin^n x \, dx \\ &= -\frac{1}{n} \cos x \sin^{n-1} x + \frac{n-1}{n} I_{n-2} \end{aligned}} \quad \ldots(1)$$

therefore, I_n is expressed in terms of I_{n-2} which is a similar integral as I_n but with a reduced power

$$I_n = \int \sin^n x \, dx \text{ and } I_{n-2} = \int \sin^{n-2} x \, dx.$$

WORKED EXAMPLE 70

Use the reduction formula of equation (1) to determine

(i) $\int \sin^2 x \, dx$

(ii) $\int \sin^3 x \, dx$

(iii) $\int \sin^4 x \, dx$, and

(iv) $\int \sin^{10} x \, dx$.

Solution 70

(i) $I_2 = \int \sin^2 x \, dx = -\frac{1}{2}\cos x \sin x + \frac{1}{2}I_0$

$I_0 = \int \sin^0 x \, dx = x$

$I_2 = \int \sin^2 x \, dx = -\frac{1}{2}\cos x \sin x + \frac{1}{2}x$

again we omitted the arbitrary constant.

(ii) $I_3 = \int \sin^3 x \, dx = -\frac{1}{3}\cos x \sin^2 x + \frac{2}{3}I_1$

$I_1 = \int \sin x \, dx = -\cos x$

$I_3 = -\frac{1}{3}\cos x \sin^2 x + \frac{2}{3}(-\cos x)$

$= -\frac{1}{3}\cos x \sin^2 x - \frac{2}{3}\cos x.$

(iii) $I_4 = \int \sin^4 x \, dx$

$= -\frac{1}{4}\cos x \sin^3 x + \frac{3}{4}I_2$

$I_2 = -\frac{1}{2}\cos x \sin x + \frac{1}{2}x$

$I_4 = -\frac{1}{4}\cos x \sin^3 x$

$\quad + \frac{3}{4}\left(-\frac{1}{2}\cos x \sin x + \frac{1}{2}x\right)$

$I_4 = -\frac{1}{4}\cos x \sin^3 x - \frac{3}{8}\cos x \sin x + \frac{3}{8}x.$

(iv) $I_{10} = -\frac{1}{10}\cos x \sin^9 x + \frac{9}{10}I_8$

$I_8 = -\frac{1}{8}\cos x \sin^7 x + \frac{7}{8}I_6$

$I_6 = -\frac{1}{6}\cos x \sin^5 x + \frac{5}{6}I_4$

$I_4 = -\frac{1}{4}\cos x \sin^3 x - \frac{3}{8}\cos x \sin x + \frac{3}{8}x$ from (iii), therefore

$I_{10} = -\frac{1}{10}\cos x \sin^9 x$

$\quad + \frac{9}{10}\left(-\frac{1}{8}\cos x \sin^7 x + \frac{7}{8}I_6\right)$

$= -\frac{1}{10}\cos x \sin^9 x - \frac{9}{80}\cos x \sin^7 x + \frac{63}{80}I_6$

$I_{10} = -\frac{1}{10}\cos x \sin^9 x - \frac{9}{80}\cos x \sin^7 x$

$\quad + \frac{63}{80}\left(-\frac{1}{6}\cos x \sin^5 x + \frac{5}{6}\right.$

$\quad \left.\left(-\frac{1}{4}\cos x \sin^3 x - \frac{3}{8}\cos x \sin x + \frac{3}{8}x\right)\right)$

$I_{10} = -\frac{1}{10}\cos x \sin^9 x - \frac{9}{80}\cos x \sin^7 x$

$\quad - \frac{63}{480}\cos x \sin^5 x - \frac{63}{80}\cdot\frac{5}{6}\cdot\frac{1}{4}\cos x \sin^3 x$

$\quad - \frac{63}{80}\cdot\frac{5}{6}\cdot\frac{3}{8}\cos x \sin x + \frac{63}{80}\cdot\frac{5}{6}\cdot\frac{3}{8}x$

$I_{10} = -\frac{1}{10}\cos x \sin^9 x - \frac{9}{80}\cos x \sin^7 x$

$\quad - \frac{21}{160}\cos x \sin^5 x - \frac{21}{128}\cos x \sin^3 x$

$\quad - \frac{63}{256}\cos x \sin x + \frac{63}{256}x$

$I_n = \int \cos^n x \, dx = \int \underset{1}{\cos x}\, \underset{2}{\cos^{n-1} x}\, dx$

$= \sin x \cos^{n-1} x$

$\quad - \int \sin x (n-1)\cos^{n-2} x (-\sin x)\, dx$

$= \sin x \cos^{n-1} x + (n-1)\int \sin^2 x \cos^{n-2} x \, dx$

$= \sin x \cos^{n-1} x$

$\quad + (n-1)\int \left(1 - \cos^2 x\right)\cos^{n-2} x \, dx$

$= \sin x \cos^{n-1} x + (n-1)\int \cos^{n-2} x \, dx$

$\quad - (n-1)\int \cos^n x \, dx$

$$I_n = \sin x \cos^{n-1} x + (n-1)I_{n-2} - (n-1)I_n$$

$$I_n(1 + n - 1) = \sin x \cos^{n-1} x + (n-1)I_{n-2}$$

$$\boxed{\begin{aligned} I_n &= \int \cos^n x \, dx \\ &= \frac{1}{n} \sin x \cos^{n-1} x + \frac{n-1}{n} I_{n-2} \end{aligned}} \quad \ldots(2)$$

which is again valid for $n \geq 2$.

WORKED EXAMPLE 71

Use equation (2) to determine

(i) I_2

(ii) I_5

(iii) I_8.

Solution 71

(i) $I_2 = \frac{1}{2} \sin x \cos x + \frac{1}{2} I_0$

$I_0 = \int \cos^0 x \, dx = x$

$I_2 = \frac{1}{2} \sin x \cos x + \frac{1}{2} x.$

(ii) $I_5 = \frac{1}{5} \sin x \cos^4 x + \frac{4}{5} I_3$

$I_3 = \frac{1}{3} \sin x \cos^2 x + \frac{2}{3} I_1$

$I_1 = \int \cos x \, dx = \sin x$

$I_5 = \frac{1}{5} \sin x \cos^4 x + \frac{4}{5} \left(\frac{1}{3} \sin x \cos^2 x + \frac{2}{3} \sin x \right)$

$I_5 = \frac{1}{5} \sin x \cos^4 x + \frac{4}{15} \sin x \cos^2 x + \frac{8}{15} \sin x.$

(iii) $I_8 = \frac{1}{8} \sin x \cos^7 x + \frac{7}{8} I_6$

$I_6 = \frac{1}{6} \sin x \cos^5 x + \frac{5}{6} I_4$

$I_4 = \frac{1}{4} \sin x \cos^3 x + \frac{3}{4} I_2$

$= \frac{1}{4} \sin x \cos^3 x + \frac{3}{4} \cdot \frac{1}{2} \sin x \cos x + \frac{3}{4} \cdot \frac{1}{2} x$

$I_6 = \frac{1}{6} \sin x \cos^5 x$

$+ \frac{5}{6} \left(\frac{1}{4} \sin x \cos^3 x + \frac{3}{8} \sin x \cos x + \frac{3}{8} x \right)$

$= \frac{1}{6} \sin x \cos^5 x + \frac{5}{24} \sin x \cos^3 x$

$+ \frac{15}{48} \sin x \cos x + \frac{15}{48} x$

$I_8 = \frac{1}{8} \sin x \cos^7 x$

$+ \frac{7}{8} \left(\frac{1}{6} \sin x \cos^5 x + \frac{5}{24} \sin x \cos^3 x \right.$

$\left. + \frac{15}{48} \sin x \cos x + \frac{15}{48} x \right)$

$I_8 = \frac{1}{8} \sin x \cos^7 x + \frac{7}{48} \sin x \cos^5 x$

$+ \frac{35}{192} \sin x \cos^3 x$

$+ \frac{105}{384} \sin x \cos x + \frac{105}{384} x.$

$$I_n = \int \tan^n x \, dx = \int \tan^{n-2} x \tan^2 x \, dx$$

$$= \int \tan^{n-2} x \left(\sec^2 x - 1 \right) dx$$

$$= \int \tan^{n-2} x \sec^2 x \, dx - \int \tan^{n-2} x \, dx$$

$$= \int \tan^{n-2} x \, d(\tan x) - I_{n-2}$$

$$\boxed{I_n = \int \tan^n x \, dx = \frac{\tan^{n-1} x}{n-1} - I_{n-2}} \quad \ldots(3)$$

which is valid for $n \geq 2$.

WORKED EXAMPLE 72

Determine

(i) $\int \tan^2 x \, dx$

(ii) $\int \tan^4 x \, dx$ and

(iii) $\int \tan^8 x \, dx$, using formula (3).

Solution 72

(i) $\int \tan^2 x \, dx = I_2 = \tan x - I_0$

$I_0 = \int dx = x$

$I_2 = \tan x - x.$

(ii) $\int \tan^4 x \, dx = I_4 = \dfrac{\tan^3 x}{3} - I_2$

$= \dfrac{\tan^3 x}{3} - \tan x + x.$

(iii) $\int \tan^8 x \, dx, = \dfrac{\tan^7 x}{7} - I_6$

$= \dfrac{\tan^7 x}{7} - \dfrac{\tan^5 x}{5} + I_4$

$I_8 = \dfrac{\tan^7 x}{7} - \dfrac{\tan^5 x}{5} + \dfrac{\tan^3 x}{3} - \dfrac{\tan x}{1} + x.$

WORKED EXAMPLE 73

If $I_n = \int \cot^n x \, dx$, prove that

$I_n = -\dfrac{1}{n-1} \cot^{n-1} x - I_{n-2},$

where $n \geq 2$, and hence find $\int \cot^5 x \, dx$.

Solution 73

$I_n = \int \cot^n x \, dx = \int \cot^{n-2} x \cot^2 x \, dx$

$= \int \cot^{n-2} x \left(\operatorname{cosec}^2 x - 1\right) dx$

$= \int \cot^{n-2} x \operatorname{cosec}^2 x \, dx - I_{n-2}$

$= -\int \cot^{n-2} x \, d(\cot x) - I_{n-2}$

$$\boxed{I_n = \int \cot^n x \, dx = -\dfrac{\cot^{n-1} x}{n-1} - I_{n-2}} \quad \ldots(4)$$

$I_5 = -\dfrac{\cot^4 x}{4} - I_3$

$I_3 = -\dfrac{\cot^2 x}{2} - I_1$

$I_1 = \int \cot x \, dx = \int \dfrac{\cos x}{\sin x} dx = \int \dfrac{d(\sin x)}{\sin x} = \ln \sin x$

$I_3 = -\dfrac{\cot^2 x}{2} - \ln \sin x$

$I_5 = -\dfrac{\cot^4 x}{4} + \dfrac{\cot^2 x}{2} + \ln \sin x.$

WORKED EXAMPLE 74

Show that a reduction formula for $I_n = \int \sec^n x \, dx$ is given by the formula,

$I_n = \dfrac{1}{n-1} \tan x \sec^{n-2} x + \dfrac{n-2}{n-1} I_{n-2}$ where $n \geq 2$, and hence find I_5.

Solution 74

$I_n = \int \sec^n x \, dx = \int \underset{2}{\sec^{n-2} x} \underset{1}{\sec^2 x} \, dx$

$= \tan x \sec^{n-2} x$

$\quad - \int \tan x (n-2) \sec^{n-3} x \sec x \tan x \, dx$

$= \tan x \sec^{n-2} x - (n-2) \int \tan^2 x \sec^{n-2} x \, dx$

but $\int \tan^2 x \sec^{n-2} x \, dx = \int \left(\sec^2 x - 1\right) \sec^{n-2} x \, dx$

$I_n = \tan x \sec^{n-2} x$

$\quad - (n-2) \int \sec^n x \, dx + (n-2) \int \sec^{n-2} x \, dx$

$I_n(1 + n - 2) = \tan x \sec^{n-2} x + (n-2) I_{n-2}$

$$\boxed{\begin{aligned} I_n &= \int \sec^n x \, dx \\ &= \frac{1}{n-1} \tan x \sec^{n-2} x + \frac{n-2}{n-1} I_{n-2} \end{aligned}} \quad \ldots(5)$$

$n = 5$

$I_5 = \frac{1}{4} \tan x \sec^3 x + \frac{3}{4} I_3$

$I_3 = \frac{1}{2} \tan x \sec x + \frac{1}{2} I_1$

$I_1 = \int \sec x \, dx = \ln|\sec x + \tan x|$

$I_5 = \frac{1}{4} \tan x \sec^3 x + \frac{3}{4} \left(\frac{1}{2} \tan x \sec x + \frac{1}{2} I_1 \right)$

$= \frac{1}{4} \tan x \sec^3 x + \frac{3}{8} \tan x \sec x$

$\quad + \frac{3}{8} \ln|\sec x + \tan x|.$

WORKED EXAMPLE 75

If $I_n = \int \cosec^n x \, dx$, show that

$I_n = -\frac{1}{n+1} \cot x \cosec^{n-2} x + \frac{n-2}{n-1} I_{n-2}$ where $n \geq 2$, hence determine I_7.

Solution 75

$I_n = \int \cosec^n x \, dx$

$= \int \underset{2}{\cosec^{n-2} x} \, \underset{1}{\cosec^2 x} \, dx$

$= -\cot x \cosec^{n-2} x - (n-2)$

$\int -\cot x \cosec^{n-3} x (-\cosec x \cot x) \, dx$

$= -\cot x \cosec^{n-2} x$

$\quad -(n-2) \int \cosec^{n-2} x \cot^2 x \, dx$

$= -\cot x \cosec^{n-2} x$

$\quad -(n-2) \int \cosec^{n-2} x \left(\cosec^2 x - 1 \right) dx$

$I_n = -\cot x \cosec^{n-2} x$

$\quad -(n-2) \int \cosec^n x \, dx + (n-2) I_{n-2}$

$I_n (1 + n - 2) = -\cot x \cosec^{n-2} x + (n-2) I_{n-2}$

$$\boxed{\begin{aligned} I_n &= \int \cosec^n x \, dx \\ &= -\frac{1}{n-1} \cot x \cosec^{n-2} x + \frac{n-2}{n-1} I_{n-2} \end{aligned}} \quad \ldots(6)$$

$I_7 = -\frac{1}{6} \cot x \cosec^5 x + \frac{5}{6} I_5$

$I_5 = -\frac{1}{4} \cot x \cosec^3 x + \frac{3}{4} I_3$

$I_3 = -\frac{1}{2} \cot x \cosec x + \frac{1}{2} I_1$

$I_1 = \ln \left| \tan \frac{1}{2} x \right|$

$I_7 = -\frac{1}{6} \cot x \cosec^5 x + \frac{5}{6} \left(-\frac{1}{4} \cot x \cosec^3 x + \frac{3}{4} I_3 \right)$

$= -\frac{1}{6} \cot x \cosec^5 x - \frac{5}{24} \cot x \cosec^3 x$

$\quad + \frac{15}{24} \left(-\frac{1}{2} \cot x \cosec x + \frac{1}{2} \ln \left| \tan \frac{1}{2} x \right| \right)$

$I_7 = -\frac{1}{6} \cot x \cosec^5 - \frac{5}{24} \cot x \cosec^3 x$

$\quad - \frac{15}{48} \cot x \cosec x + \frac{15}{48} \ln \left| \tan \frac{1}{2} x \right|.$

WORKED EXAMPLE 76

Determine reduction formulae for the following:

(i) $I_n = \int \sinh^n x \, dx$

(ii) $I_n = \int \cosh^n x \, dx$

(iii) $I_n = \int \tanh^n x \, dx$

Solution 76

(i) $I_n = \int \sinh^n x \, dx = \int \underset{2}{\sinh^{n-1} x} \underset{1}{\sinh x} \, dx$

$= \cosh x \sinh^{n-1} x$

$\quad - \int \cosh x (n-1) \sinh^{n-2} x \cosh x \, dx$

$I_n = \cosh x \sinh^{n-1} x$

$\quad - \int \cosh^2 x (n-1) \sinh^{n-2} x \, dx$

$= \cosh x \sinh^{n-1} x$

$\quad - (n-1) \int \left(1 + \sinh^2 x\right) \sinh^{n-2} x \, dx$

$= \cosh x \sinh^{n-1} x$

$\quad - (n-1) \int \sinh^{n-2} x \, dx$

$\quad - (n-1) \int \sinh^n x \, dx$

$I_n(1 + n - 1) = \cosh x \sinh^{n-1} x - (n-1) I_{n-2}$

$\boxed{n I_n = \cosh x \sinh^{n-1} x - (n-1) I_{n-2}.}$

(ii) $I_n = \int \cosh^n x \, dx = \int \underset{2}{\cosh^{n-1} x} \underset{1}{\cosh x} \, dx$

$= \sinh x \cosh^{n-1} x$

$\quad - \int \sinh x (n-1) \cosh^{n-2} x \sinh x \, dx$

$= \sinh x \cosh^{n-1} x$

$\quad - \int \sinh^2 x (n-1) \cosh^{n-2} x \, dx$

$= \sinh x \cosh^{n-1} x$

$\quad - \int \left(\cosh^2 x - 1\right)(n-1) \cosh^{n-2} x \, dx$

$= \sinh x \cosh^{n-1} x - (n-1) \int \cosh^n x \, dx$

$\quad + (n-1) \int \cosh^{n-2} x \, dx$

$I_n(1 + n - 1) = \sinh x \cosh^{n-1} x + (n-1) I_{n-2}$

$\boxed{n I_n = \sinh x \cosh^{n-1} x + (n-1) I_{n-2}}$

(iii) $I_n = \int \tanh^n x \, dx$

$= \int \tanh^{n-2} x \tanh^2 x \, dx$

$= \int \tanh^{n-2} x \left(1 - \text{sech}^2 x\right) dx$

$= \int \underset{2}{\tanh^{n-2} x} \underset{1}{dx} - \int \underset{2}{\tanh^{n-2} x} \underset{1}{\text{sech}^2 x} \, dx$

$= I_{n-2} - \tanh x \tanh^{n-2} x$

$\quad + \int \tanh x (n-2) \tanh^{n-3} x \times \text{sech}^2 x \, dx$

$= I_{n-2} - \tanh x \tanh^{n-2} x$

$\quad + (n-2) \int \tanh^{n-2} x \left(1 - \tanh^2 x\right) dx$

$= I_{n-2} - \tanh x \tanh^{n-2} x$

$\quad + (n+2) \int \tanh^{n-2} x \, dx - (n-2) I_n$

$I_n(1 + n - 2) = I_{n-2} - \tanh x \tanh^{n-2} x$

$\quad + (n-2) I_{n-2}$

$\boxed{I_n(n-1) = (n-1) I_{n-2} - \tanh x \tanh^{n-2} x.}$

Definite Integrals

If $I_n = \int_0^{\frac{\pi}{4}} \sin^n x \, dx$

$I_n = \int_0^{\frac{\pi}{4}} \sin^n x \, dx$

$= \left[-\frac{1}{n} \cos x \sin^{n-1} x\right]_0^{\frac{\pi}{4}} + \frac{n-1}{n} I_{n-2}$

$= -\frac{1}{n} \cos \frac{\pi}{4} \sin^{n-1} \frac{\pi}{4} + \frac{n-1}{n} I_{n-2}$

$= -\frac{1}{n} \frac{1}{\sqrt{2}} \left(\frac{1}{\sqrt{2}}\right)^{n-1} + \frac{n-1}{n} I_{n-2}$

58 — GCE A level

$$I_{n-2} = \int_0^{\frac{\pi}{4}} \sin^{n-2} x \, dx$$

$$= \left[-\frac{1}{n-2} \cos x \sin^{n-3} x\right]_0^{\frac{\pi}{4}} + \frac{n-3}{n-2} I_{n-4}$$

$$= -\frac{1}{n-2} \cdot \frac{1}{\sqrt{2}} \left(\frac{1}{\sqrt{2}}\right)^{n-3} + \frac{n-3}{n-2} I_{n-4}$$

$$I_{n-4} = \int_0^{\frac{\pi}{4}} \sin^{n-4} x \, dx$$

$$= \left[-\frac{1}{n-4} \cos x \sin^{n-5} x\right]_0^{\frac{\pi}{4}} + \frac{n-5}{n-4} I_{n-6}$$

$$= -\frac{1}{n-4} \cdot \frac{1}{\sqrt{2}} \left(\frac{1}{\sqrt{2}}\right)^{n-5} + \frac{n-5}{n-4} I_{n-6}$$

$$I_n = -\frac{1}{n}\left(\frac{1}{\sqrt{2}}\right)^n + \frac{n-1}{n}\left(-\frac{1}{n-2}\right)\left(\frac{1}{\sqrt{2}}\right)^{n-2}$$

$$+ \left(\frac{n-3}{n-2}\right)\left(-\frac{1}{n-4}\right)\left(\frac{1}{\sqrt{2}}\right)^{n-4}$$

$$+ \frac{n-5}{n-4} I_{n-6} \text{ and so on.}$$

If $I_n = \int_0^{\frac{\pi}{2}} \sin^n x \, dx = \left[-\frac{1}{n} \cos x \sin^{n-1} x\right]_0^{\frac{\pi}{2}}$

$$+ \frac{n-1}{n} I_{n-2} = \frac{n-1}{n} I_{n-2}$$

$$I_{n-2} = \int_0^{\frac{\pi}{2}} \sin^{n-2} x \, dx = \left[-\frac{1}{n-2} \cos x \sin^{n-3} x\right]_0^{\frac{\pi}{2}}$$

$$+ \frac{n-3}{n-2} I_{n-4} = \left[\frac{n-3}{n-2} I_{n-4}\right]$$

$$I_{n-4} = \int_0^{\frac{\pi}{2}} \sin^{n-4} x \, dx$$

$$= \left[-\frac{1}{n-4} \cos x \sin^{n-5} x\right]_0^{\frac{\pi}{2}} + \frac{n-5}{n-4} I_{n-6}$$

$$I_{n-4} = \frac{n-5}{n-4} I_{n-6}$$

$$I_n = \frac{n-1}{n} \cdot \frac{n-3}{n-2} \cdot \frac{n-5}{n-4} \cdot I_{n-6}$$

If n is even

$$I_n = \frac{n-1}{n} \cdot \frac{n-3}{n-2} \cdot \frac{n-5}{n-4} \cdot \frac{n-7}{n-6} \cdots \frac{3}{4} \cdot \frac{1}{2} I_0$$

$$I_0 = \int_0^{\frac{\pi}{2}} dx = \frac{\pi}{2}$$

$$\boxed{I_n = \frac{n-1}{n} \cdot \frac{n-3}{n-2} \cdot \frac{n-5}{n-4} \cdots \frac{3}{4} \cdot \frac{1}{2} \cdot \frac{\pi}{2}} \quad \ldots(1)$$

If n is odd

$$\boxed{I_n = \frac{n-1}{n} \cdot \frac{n-3}{n-2} \cdot \frac{n-5}{n-4} \cdots \frac{4}{5} \cdot \frac{2}{3} \cdot I_1} \quad \ldots(2)$$

$$I_1 = \int_0^{\frac{\pi}{2}} \sin x \, dx = \left[-\cos x\right]_0^{\frac{\pi}{2}} = -\cos \frac{\pi}{2} + \cos 0 = 1$$

WORKED EXAMPLE 77

Evaluate

(a) (i) $\int_0^{\frac{\pi}{2}} \sin^2 x \, dx$

(ii) $\int_0^{\frac{\pi}{2}} \sin^4 x \, dx$

(iii) $\int_0^{\frac{\pi}{2}} \sin^6 x \, dx$

(b) (i) $\int_0^{\frac{\pi}{2}} \sin^3 x \, dx$

(ii) $\int_0^{\frac{\pi}{2}} \sin^5 x \, dx$

(iii) $\int_0^{\frac{\pi}{2}} \sin^7 x \, dx$

Solution 77

(a) (i) $\int_0^{\pi/2} \sin^2 x \, dx = \frac{1}{2} \frac{\pi}{2} = \frac{\pi}{4}$

(ii) $\int_0^{\pi/2} \sin^4 x \, dx = \frac{3}{4} \cdot \frac{1}{2} \frac{\pi}{2} = \frac{3\pi}{16}$

(iii) $\int_0^{\pi/2} \sin^6 x \, dx = \frac{5}{6} \cdot \frac{3}{4} \cdot \frac{1}{2} \cdot \frac{\pi}{2} = \frac{15\pi}{96}$

applying equation (1) for n even

(b) (i) $\int_0^{\pi/2} \sin^3 x \, dx = \frac{2}{3} \cdot 1 = \frac{2}{3}$

(ii) $\int_0^{\pi/2} \sin^5 x \, dx = \frac{4}{5} \cdot \frac{2}{3} \cdot 1 = \frac{8}{15}$

(iii) $\int_0^{\pi/2} \sin^7 x \, dx = \frac{6}{7} \cdot \frac{4}{5} \cdot \frac{2}{3} \cdot 1 = \frac{48}{105}$

applying equation (2) for n odd.

If $I_n = \int_0^{\pi/2} \cos^n x \, dx = \left[\frac{1}{n} \sin x \cos^{n-1} x \right]_0^{\pi/2}$

$+ \frac{n-1}{n} I_{n-2} = \frac{n-1}{n} I_{n-2}$

If n is even

$I_n = \frac{n-1}{n} \cdot \frac{n-3}{n-2} \cdots \frac{3}{4} \cdot \frac{1}{2} \cdot I_0$

$I_0 = \int_0^{\pi/2} \cos^0 x \, dx = \frac{\pi}{2}$

$\boxed{I_n = \frac{n-1}{n} \cdot \frac{n-3}{n-2} \cdots \frac{3}{4} \cdot \frac{1}{2} \cdot \frac{\pi}{2}}$ ...(3)

If n is odd

$I_n = \frac{n-1}{n} \cdot \frac{n-3}{n-2} \cdots \frac{4}{5} \cdot \frac{2}{3} \cdot I_1$

$I_1 = \int_0^{\pi/2} \cos x \, dx = \left[\sin x \right]_0^{\pi/2} = 1$

$\boxed{I_n = \frac{n-1}{n} \cdot \frac{n-3}{n-2} \cdots \frac{4}{5} \cdot \frac{2}{3} \cdot 1}$ ...(4)

WORKED EXAMPLE 78

Evaluate

(a) (i) $\int_0^{\pi/2} \cos^4 x \, dx$

(ii) $\int_0^{\pi/2} \cos^8 x \, dx$

(b) (i) $\int_0^{\pi/2} \cos^5 x \, dx$

(ii) $\int_0^{\pi/2} \cos^9 x \, dx$.

Solution 78

(a) (i) $\int_0^{\pi/2} \cos^4 x \, dx = \frac{3}{4} \cdot \frac{1}{2} \cdot \frac{\pi}{2} = \frac{3\pi}{16}$

(ii) $\int_0^{\pi/2} \cos^8 x \, dx = \frac{7}{8} \cdot \frac{5}{6} \cdot \frac{3}{4} \cdot \frac{1}{2} \cdot \frac{\pi}{2} = \frac{105\pi}{768}$

(b) (i) $\int_0^{\pi/2} \cos^5 x \, dx = \frac{4}{5} \cdot \frac{2}{3} \cdot 1 = \frac{8}{15}$

(ii) $\int_0^{\pi/2} \cos^9 x \, dx = \frac{8}{9} \cdot \frac{6}{7} \cdot \frac{4}{5} \cdot \frac{2}{3} \cdot 1 = \frac{384}{945}.$

Reduction Formulae

If $I_n = \displaystyle\int_0^\infty \dfrac{1}{(1+x^2)^n}\,dx$, where $n \in Z^+$, show that

$$I_{n-1} = \dfrac{2n-2}{2n-3} I_n, \text{ for } n \geq 2.$$

Solution

$$I_n = \int_0^\infty \dfrac{1}{(1+x^2)^n}\,dx,$$

$$I_{n-1} = \int_0^\infty \dfrac{1}{(1+x^2)^{n-1}}\,dx$$

$$I_{n-1} = \int_0^\infty \dfrac{1}{(1+x^2)^{n-1}}\,dx$$

$$= \int_0^\infty \underbrace{(1+x^2)^{-n+1}}_{2}\,\underbrace{dx}_{1}$$

$$= \left[x(1+x^2)^{-n+1}\right]_0^\infty$$

$$- \int_0^\infty x(-n+1)(1+x^2)^{-n}(2x)\,dx$$

$$= +2(n-1)\int_0^\infty x^2(1+x^2)^{-n}\,dx$$

$$= 2(n-1)\int_0^\infty (x^2+1-1)(1+x^2)^{-n}\,dx$$

$$= 2(n-1)\int_0^\infty \left[(1+x^2)^{-n+1} - (1+x^2)^{-n}\right]dx$$

$$I_{n-1} = 2(n-1)\int_0^\infty \dfrac{1}{(1+x^2)^{n-1}}\,dx$$

$$- 2(n-1)\int_0^\infty \dfrac{1}{(1+x^2)^{-n}}\,dx$$

$$I_{n-1} = 2(n-1)I_{n-1} - 2(n-1)I_n$$

$$I_{n-1}(1-2n+2) = -2(n-1)I_n$$

$$\boxed{I_{n-1} = \dfrac{2n-2}{2n-3} I_n}$$

Hence evaluate I_3, I_5 and I_7.

$$I_n = \dfrac{2n-3}{2n-2} I_{n-1}$$

$$n=3,\ I_3 = \dfrac{3}{4} I_2,\ I_2 = \dfrac{1}{2} I_1$$

$$I_1 = \int_0^\infty \dfrac{1}{(1+x^2)}\,dx = \left[\tan^{-1} x\right]_0^\infty = \dfrac{\pi}{2}$$

$$I_3 = \dfrac{3}{4}\cdot\dfrac{1}{2}\cdot\dfrac{\pi}{2} = \dfrac{3\pi}{16}$$

$n = 5$

$$I_5 = \dfrac{7}{8} I_4 = \dfrac{7}{8}\cdot\dfrac{5}{6}\cdot I_3$$

$$I_5 = \dfrac{7}{8}\cdot\dfrac{5}{6}\cdot\dfrac{3\pi}{16} = \dfrac{105}{768}\pi$$

$n = 7$

$$I_7 = \dfrac{11}{12} I_6 = \dfrac{11}{12}\cdot\dfrac{9}{10} I_5$$

$$I_7 = \dfrac{11}{12}\cdot\dfrac{9}{10}\cdot\dfrac{105}{768}\pi = \dfrac{10395}{84490}\pi.$$

WORKED EXAMPLE 79

Show

$$\int x^p (\log_e x)^q\,dx = \dfrac{x^{p+1}}{p+1}(\log_e x)^q$$

$$- \dfrac{q}{p+1}\int x^p (\log_e x)^{q-1}\,dx$$

hence evaluate $\displaystyle\int_1^2 x^3 (\log_e x)^3\,dx.$

Solution 79

To show

$$\int x^p (\log_e x)^q \, dx = \frac{x^{p+1}}{p+1} (\log_e x)^q$$
$$- \frac{q}{q+1} \int x^p (\log_e x)^{q-1} \, dx$$

$$\int x^p (\log_e x)^q \, dx = \frac{x^{p+1}}{p+1} (\log_e x)^q$$

$$- \int \frac{x^{p+1}}{p+1} q (\log_e x)^{q-1} \frac{1}{x} \, dx$$

$$= \frac{x^{p+1}}{p+1} (\log_e x)^q$$

$$- \int \frac{x^p}{p+1} q (\log_e x)^{q-1} \, dx$$

$$= \frac{x^{p+1}}{p+1} (\log_e x)^q$$

$$- \frac{q}{p+1} \int x^p (\log_e x)^{q-1} \, dx$$

$p = 3, q = 3$

$$\int_1^2 x^3 (\log_e x)^3 \, dx = \left[\frac{x^4}{4} (\log_e x)^3 \right]_1^2$$

$$- \frac{3}{4} \int_1^2 x^3 (\log_e x)^2 \, dx$$

$$\int_1^2 x^3 (\log_e x)^2 \, dx = \left[\frac{x^4}{4} (\log_e x)^2 \right]_1^2$$

$$- \frac{2}{4} \int_1^2 x^3 (\log_e x) \, dx$$

$$\int_1^2 x^3 (\log_e x) \, dx = \left[\frac{x^4}{4} (\log_e x) \right]_1^2 - \frac{1}{4} \int_1^2 x^3 \, dx$$

$$\int_1^2 x^3 \, dx = \left[\frac{x^4}{4} \right]_1^2 = 4 - \frac{1}{4} = \frac{15}{4}$$

$$\therefore \int_1^2 x^3 (\log_e x)^3 \, dx = 4 (\log_e 2)^3$$

$$- \frac{3}{4} \left[4 (\log_e 2)^2 - \frac{1}{2} \left(4 \log_e 2 - \frac{1}{4} \times \frac{15}{4} \right) \right]$$

$$= 4 (\log_e 2)^3 - 3 (\log_e 2)^2 + \frac{3}{8}$$

$$\times 4 \log_e 2 - \frac{3}{8} \times \frac{1}{4} \times \frac{15}{4}$$

$$= 1.33 - 1.44 + 1.04 - 0.352 = 0.578$$

Exercises 8

1. Show that

$$\int \cos^n x \, dx = \frac{1}{n} \cos^{n-1} x \sin x$$
$$+ \frac{n-1}{n} \int \cos^{n-2} x \, dx$$

and hence find

(i) $\int \cos^3 x \, dx$

(ii) $\int \cos^9 x \, dx$

(iii) $\int_0^{\frac{\pi}{2}} \cos^5 x \, dx$.

2. If $I_n = \int x^n e^x \, dx$, find a reduction formula and hence determine I_6 between the limits $x = 0$ and $x = 1$.

3. If $I_n = \int x^n \cos x \, dx$, find a reduction formula and hence determine I_4.

4. Find a reduction formula for $I_n = \int x (\ln x)^n \, dx$

evaluate $\int_1^2 x (\ln x)^5 dx$.

5. Find a reduction formula for $I_n = \int x^n \ln x \, dx$

evaluate $\int_1^2 x^{20} \ln x \, dx$.

6. Find a reduction formula for
$$I_n = \int \left(1+x^2\right)^n dx \text{ hence find } I_4.$$

7. Find a reduction formula for
$$I_n = \int \left(1+x^2\right)^{-n} dx \text{ hence find } I_3.$$

8. Find a reduction formula for
$$I_n = \int \sin^n 2x \, dx \text{ hence find } I_3.$$

9. Find a reduction formula for
$$I_n = \int \tan^n 3x \, dx \text{ hence find } I_5.$$

10. Find a reduction formula for
$$I_n = \int x^n \cosh x \, dx.$$

11. If $I_{m,n} = \int \sin^m x \cos^n x \, dx$, prove that
$$I_{m,n} = \int \sin^{m-1} x \frac{d}{dx}\left(\frac{-\cos^{n+1} x}{n+1}\right) dx$$
$$= -\frac{\sin^{m-1} x \cos^{n+1} x}{n+1}$$
$$+ \frac{m-1}{n+1}\left(I_{m-2,\,n} - I_{m,\,n}\right)$$

hence show that
$$I_{m,\,n} = -\frac{\sin^{m-1} x \cos^{n+1} x}{m+n} + \frac{m-1}{m+n} I_{m-2,\,n}.$$

12. If $I_n = \int_0^{\frac{\pi}{2}} x \sin^n x \, dx$, and $n > 1$, prove that $I_n = \frac{n-1}{n} I_{n-2} + \frac{1}{n^2}$, hence find the value of I_3 and show that $I_5 = \frac{149}{225}$.

9

Approximate Numerical Integration

Derivation of the Trapezoidal Rule

Consider the function of $y = e^x$ form $x = 0$ to $x = 1$.

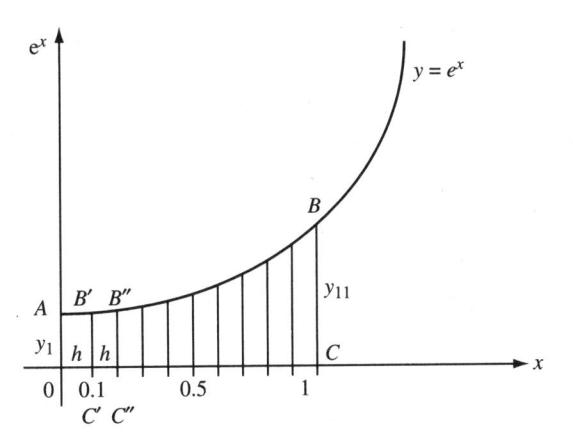

Fig. 7-I/27

The boundary of the area is $OABC$. The base width OC is divided into equal intervals, either <u>odd</u> or <u>even</u>.

For convenience, let us divide OC into ten equal intervals of h width or eleven ordinates, 10 intervals correspond to $10 + 1 = 11$ ordinates. Evaluate e^x to four decimal places.

x	$y = e^x$
0	1
0.1	1.1052
0.2	1.2214
0.3	1.3499
0.4	1.4918
0.5	1.6487
0.6	1.8221

x	$y = e^x$
0.7	2.0138
0.8	2.2255
0.9	2.4596
1.0	2.7183

The area of the trapezium $OAB'C'$ is given by the formula where $OA = y_1$, $B'C' = y_2$, $OC' = h = 0.1$ the area of the trapezium $B'B''C''C'$ is given by $\frac{1}{2}(y_2 + y_3)h$ where $B''C'' = y_3$ and so on for the remaining eight intervals.

Summing up all these ten interval areas, we have

$$\frac{1}{2}(y_1 + y_2)h + \frac{1}{2}(y_2 + y_3)h$$
$$+ \frac{1}{2}(y_3 + y_4)h + \ldots + \frac{1}{2}(y_{10} + y_{11})h$$

$$\int y \, dx \approx \frac{h}{2}[y_1 + y_{11} + 2(y_2 + y_3 + y_4 + y_5 + y_6 + y_7 + y_8 + y_9 + y_{10})].$$

In general, the Trapezium Rule is given

$$\int_{x_0}^{x_n} f(x) \, dx$$
$$\approx \frac{h}{2}[f_0 + 2(f_1 + f_2 + \ldots + f_{n-1}) + f_n]$$

$f_r = f(x_r)$, where $x_r = x_0 + rh$.

Derivation of the Mid-Ordinate Rule

Considering again the same example of the exponential function $y = e^x$ from $x = 0$ to $x = 1$ for intervals and base width $h = 0.1$.

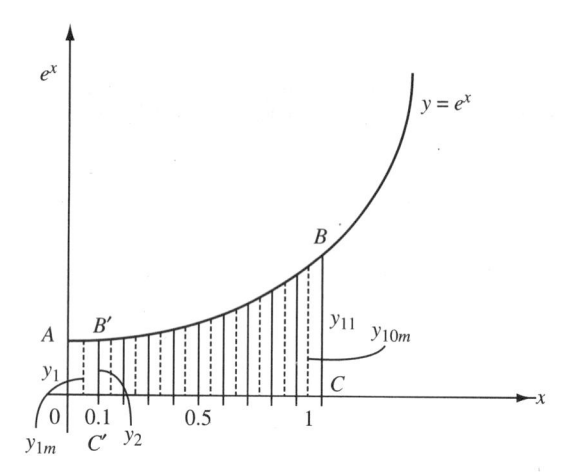

Fig. 7-I/28

There are ten intervals, eleven ordinates, and ten mid-ordinates (shown dotted).

To find the first mid-ordinate, $y_{1m} = \frac{1}{2}(y_1 + y_2)$, the second mid-ordinate, $y_{2m} = \frac{1}{2}(y_2 + y_3)$, the third $y_{3m} = \frac{1}{2}(y_3 + y_4)$ and so on to

$y_{10m} = \frac{1}{2}(y_{10} + y_{11})$. The approximate area of $AB'C'O$ = area of rectangle = $y_{1m}h$. The total area is the sum of the ten rectangles.

Area = $y_{1m}h + y_{2m}h + y_{3m}h + \ldots + y_{10m}h$

Area = $h(y_{1m} + y_{2m} + y_{3m} + \ldots + y_{10m})$

Area = the sum of the mid-ordinates times the interval width (h).

Certain definite integrals which cannot be evaluated by conventional integration, they can be evaluated by the numerical method of the trapezoidal and mid-ordinate rules.

Such integrals are $\int_0^{\frac{\pi}{4}} \sqrt{\sin x}\, dx$, $\int_0^1 e^{x^2}\, dx$.

WORKED EXAMPLE 80

Evaluate approximately to four decimal places the integral $\int_0^1 e^x\, dx$ by using

(i) the trapezoidal rule with ten equal intervals

(ii) the mid-ordinate rule with ten equal intervals.

What is the exact value of the integral?

What is the percentage error?

Solution 80

(i) $\int_0^1 e^x\, dx \approx \frac{h}{2}[y_1 + y_{11} + 2(y_2 + y_3 + y_4 + y_5 + y_6 + y_7 + y_8 + y_9 + y_{10})]$

$h = \frac{1-0}{10} = \frac{\text{upper limit} - \text{lower limit}}{\text{number of intervals}} = 0.1$

x	y
0	1
0.1	1.1052
0.2	1.2214
0.3	1.3499
0.4	1.4918
0.5	1.6487
0.6	1.8221
0.7	2.0138
0.8	2.2255
0.9	2.4596
1.0	2.7183

$\int_0^1 e^x\, dx \approx \frac{0.1}{2}[1 + 2.7183 + 2(1.1052$
$+ 1.2214 + 1.3499 + 1.4918$
$+ 1.6487 + 1.8221 + 2.0138$
$+ 2.2255 + 2.4596)]$

$= 1.7197.$

(ii) $\int_0^1 e^x \, dx \approx h(y_{1m} + y_{2m} + y_{3m} + \ldots + y_{10m})$

$= 0.1(1.0526 + 1.1633 + 1.2857$
$\quad + 1.4209 + 1.5703 + 1.7354 + 1.9180$
$\quad + 2.1197 + 2.3426 + 2.5890)$
$= 1.7198.$

(iii) $\int_0^1 e^x \, dx = \left[e^x\right]_0^1 = e^1 - 1 = 1.7183.$

In 1.7183 the error is $1.7198 - 1.7183$ or 0.0015, in 100, the error will be $\dfrac{0.15}{1.7183} = 0.087\%$, less than 0.1% error.

x	$y = \log x$
1.2	0.0792
1.3	0.1139
1.4	0.1461
1.5	0.1761
1.6	0.2041
1.7	0.2304
1.8	0.2553
1.9	0.2788
2.0	0.3010

WORKED EXAMPLE 81

Evaluate approximately to three decimal places using the trapezoidal rule, the following definite integrals:

(i) $\int_1^2 \log_{10} x \, dx$

(ii) $\int_0^{\pi/4} \sqrt{\sin x} \, dx$

using 10 intervals or 11 ordinates.

Solution 81

(i) $\int_1^2 \log_{10} x \, dx \approx \dfrac{0.1}{2}[0 + 0.3010$
$\quad + 2(0.0414 + 0.0792 + 0.1139$
$\quad + 0.1461 + 0.1761 + 0.2041$
$\quad + 0.2304 + 0.2553 + 0.2788)]$
$= 0.168$

where $h = \dfrac{2-1}{10} = 0.1$

x	$y = \log x$
1	0
1.1	0.0414

(ii) $\int_0^{\pi/4} \sqrt{\sin x} \, dx \approx \dfrac{\pi}{2 \times 40}[0 + 0.8409$

$\quad + 2(0.2801 + 0.3955 + 0.4832$
$\quad + 0.5559 + 0.6186 + 0.6738$
$\quad + 0.7228 + 0.7667 + 0.8059)]$
$= 0.449$

where $h = \dfrac{\pi/4}{10} = \dfrac{\pi}{40}$.

x	$y = \sqrt{\sin x}$
0	0
$\dfrac{\pi}{40}$	0.2801
$\dfrac{2\pi}{40}$	0.3955
$\dfrac{3\pi}{40}$	0.4832
$\dfrac{4\pi}{40}$	0.5559
$\dfrac{5\pi}{40}$	0.6186
$\dfrac{6\pi}{40}$	0.6738
$\dfrac{7\pi}{40}$	0.7228

x	$y = \sqrt{\sin x}$
$\dfrac{8\pi}{40}$	0.7667
$\dfrac{9\pi}{40}$	0.8059
$\dfrac{10\pi}{40}$	0.8409

Simpson's Rule

Let ABC be an arc of the curve $y = f(x)$ intersecting the y-axis at B and that AA', CC' are ordinates at $x = -h$ and $x = h$.

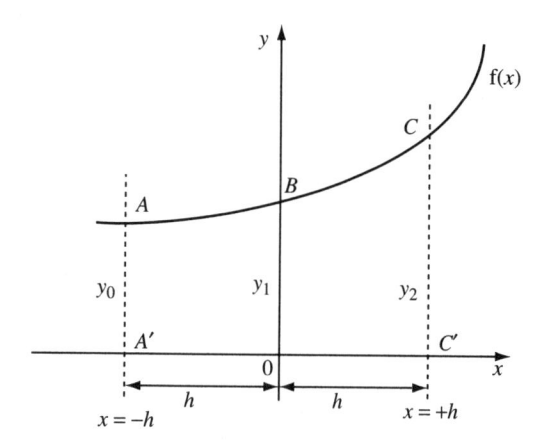

Fig. 7-I/29

$AA' = y_0$, $BO = y_1$ and $CC' = y_2$.

The equation $y = ax^2 + bx + c$ contains three constants a, b and c and passes through the points A, B and C, assuming h to be small, the area below this curve will be a good approximation to the area below the curve ABC shown in the diagram.

$$\int_{-h}^{h} f(x)\, dx = \int_{-h}^{h} (ax^2 + bx + c)\, dx$$

$$\approx \left[\frac{1}{3}ax^3 + \frac{1}{2}bx^2 + cx \right]_{-h}^{h}$$

$$= \frac{1}{3}ah^3 + \frac{1}{2}bh^2 + ch - \frac{1}{3}a(-h)^3$$

$$- \frac{1}{2}b(-h)^2 - c(-h) = \frac{2}{3}ah^3 + 2ch.$$

If the curve $ax^2 + bx + c$ passes through the points A, B and C, since $y = y_0$ when $x = -h$, $y = y_1$ when $x = 0$ and $y = y_2$ when $x = h$,

$$y_0 = ah^2 - bh + c \qquad \ldots(1)$$

$$y_1 = c \qquad \ldots(2)$$

$$y_2 = ah^2 + bh + c. \qquad \ldots(3)$$

Adding (1) and (3)

$$y_0 + y_2 = 2ah^2 + 2c$$

and since $c = y_1$

$$y_0 + y_2 = 2ah^2 + 2y_1 \qquad \ldots(4)$$

$$2ah^2 = y_0 + y_2 - 2y_1 \qquad \ldots(5)$$

substituting (4) and (5) in

$$\int_{-h}^{h} f(x)\, dx = \frac{2}{3}ah^3 + 2ch$$

$$= 2ah^2 \cdot \frac{h}{3} + 2ch$$

$$= (y_0 + y_2 - 2y_1)\frac{h}{3} + 2hy_1$$

$$\int_{-h}^{h} f(x)\, dx = \frac{h}{3}(y_0 + 4y_1 + y_2)$$

this is the approximate area of a strip of width $2h$ when an arc of the curve $y = f(x)$ is replaced by an arc of $y = ax^2 + bx + c$ which passes through three points on the given curve. Extending this theory to more such strips as in Fig. 7-I/30

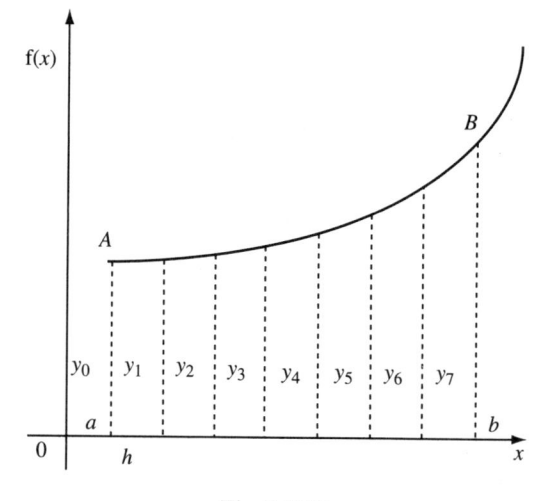

Fig. 7-I/30

$$\int_a^b y\,dx = \text{area } ABba$$

$$\approx \frac{h}{3}(y_0 + 4y_1 + y_2) + \frac{h}{3}(y_2 + 4y_3 + y_4)$$

$$+ \frac{h}{3}(y_4 + 4y_5 + y_6) + \frac{h}{3}(y_6 + 4y_7 + y_8)$$

$$\int_a^b y\,dx \approx \frac{h}{3}[y_0 + 4y_1 + 2y_0 + 4y_3$$

$$+ 2y_4 + 4y_5 + 2y_6 + 4y_7 + y_8].$$

In a similar way, if we divide the area into an even number $2n$ of strips,

$$\int_a^b y\,dx$$

$$\approx \frac{h}{3}[y_0 + y_{2n} + 4(y_1 + y_3 + y_5 + \ldots + y_{2n-1})$$

$$+ 2(y_2 + y_4 + \ldots + y_{2n-2})]$$

The number of intervals n, must be even, giving an odd number of ordinates

$$\int_{x_0}^{x_n} f(x)\,dx$$

$$\approx \frac{1}{3}h[f_0 + f_n + 4(f_1 + f_3 + \ldots + f_{n-1})$$

$$+ 2(f_2 + f_4 + \ldots + f_{n-2})]$$

WORKED EXAMPLE 82

Use Simpson's rule with eleven ordinates to find approximate value for

$$I = \int_0^1 \frac{1}{1+x^2}\,dx, \text{ to four decimal places.}$$

What is the exact value of the integral I? Use also the trapezoidal rule for eleven ordinates and comment on which rule is better.

Solution 82

x	$\frac{1}{1+x^2}$
0	1
0.1	0.990
0.2	0.962
0.3	0.917
0.4	0.862
0.5	0.800
0.6	0.735
0.7	0.671
0.8	0.610
0.9	0.553
1.0	0.500

$$h = \frac{1-0}{10} = 0.1$$

$$\int_0^1 \frac{1}{1+x^2}\,dx$$

$$\approx \frac{0.1}{3}[1 + 0.500 + 4(0.990 + 0.917 + 0.800$$

$$+ 0.671 + 0.553)$$

$$+ 2(0.962 + 0.862 + 0.735 + 0.610)]$$

$$= \frac{0.1}{3}(1.5 + 15.724 + 6.338) = 0.7854$$

$$\int_0^1 \frac{1}{1+x^2}\,dx = \left[\tan^{-1} x\right]_0^1 = \frac{\pi}{4} = 0.7854$$

$$\int_0^1 \frac{1}{1+x^2}\,dx \approx \frac{0.1}{2}[1 + 0.500 + 2(0.990 + 0.962$$

$$+ 0.917 + 0.862 + 0.800 + 0.735$$

$$+ 0.671 + 0.610 + 0.553)]$$

$$= 0.785$$

Simpson's rule seems to be more accurate.

WORKED EXAMPLE 83

(a) Evaluate exactly the definite integral $\int_0^{\frac{\pi}{2}} \sin x \, dx$.

(b) Find approximately to three significant figures the area $\int_0^{\frac{\pi}{2}} \sin x \, dx$ using

Simpson's Rule

(i) with five ordinates
(ii) seven ordinates
(iii) three ordinates.

Solution 83

(a) $\int_0^{\frac{\pi}{2}} \sin x \, dx = \left[-\cos x \right]_0^{\frac{\pi}{2}}$

$= -\cos \frac{\pi}{2} + \cos 0 = 1$ square unit

(b) (i) $\int_0^{\frac{\pi}{2}} \sin x \, dx = \frac{1}{3}\left[y_1 + y_5 + 4(y_2 + y_4) + 2y_3 \right]$

where $h = \frac{\frac{\pi}{2}}{4} = \frac{\pi}{8}$

x	y
0	0
$\frac{\pi}{8}$	0.383
$\frac{\pi}{4}$	0.707
$\frac{3\pi}{8}$	0.924
$\frac{\pi}{2}$	1

$\int_0^{\frac{\pi}{2}} \sin x \, dx \approx \frac{\frac{\pi}{8}}{3}\left[0 + 1 + 4(0.383 + 0.924) \right.$

$\left. + 2(0.707) \right]$

$= \frac{\pi}{24}[7.642] = 1.000335461$

≈ 1.00

(ii) $\int_0^{\frac{\pi}{2}} \sin x \, dx \approx \frac{\frac{\pi}{12}}{3}[0 + 1.00$

$+ 4(0.259 + 0.707 + 0.966)$

$+ 2(0.5 + 0.866)]$

$= \frac{\pi}{36}(1 + 7.728 + 2.732)$

$= 1.000073661$

≈ 1.00

x	$\sin x$
0	0
$\frac{\pi}{12}$	0.259
$\frac{2\pi}{12}$	0.500
$\frac{3\pi}{12}$	0.707
$\frac{4\pi}{12}$	0.866
$\frac{5\pi}{12}$	0.966
$\frac{\pi}{2}$	1.00

(iii) $\int_0^{\frac{\pi}{2}} \sin x \, dx \approx \frac{\frac{\pi}{4}}{3}\left[0 + 1 + 4 \times 0.707 \right]$

$= 1.002168056$

≈ 1.00

x	$\sin x$
0	0
$\frac{\pi}{4}$	0.707
$\frac{\pi}{2}$	1

It is observed that the more ordinates we consider the closer to the exact answer.

Exercises 9

1. A curve whose equation is $y = f(x)$ passes through the points defined in the following table

x	y
1	5
2	9
3	12
4	16
5	19
6	23
7	26
8	30
9	33

   Use Simpson's Rule, with nine ordinates to evaluate $\int_1^9 f(x)\,dx$ and hence determine approximately the average ordinate in the range $1 \leq x \leq 9$.

   (Ans. 154.7 square units, 19.3 units)

2. A curve whose equation is $y = f(x)$ passes through the points defined in the following table

x	y
0.1	25
0.2	29.6
0.3	35.9
0.4	42
0.5	46
0.6	52

   Use the trapezoidal rule with six ordinates to evaluate $\int_{0.1}^{0.6} y\,dx$. Explain why it is difficult to evaluate the area using Simpson's Rule.

   (Ans. 19.2 square units)

3. Use Simpson's Rule with eight intervals to find an approximate value of
   $$\int_{0.1}^{0.9} e^{x^2}\,dx$$
   (Ans. 1.12 square units).

4. Use Simpson's Rule with four intervals to evaluate an approximate value of
   $$\int_0^\pi x \cos x\,dx$$
   correct to three decimal places.

   (Ans. 1.990 square units)

5. The table below gives values of $f(x) = \frac{1}{5}\sqrt{25 - x^2}$ correct to three decimal places for values of x between 0 and 5, at intervals of 1.

x	$f(x)$
0	1.000
1.0	0.980
2.0	0.917
3.0	0.800
4.0	0.600
5.0	0.000

   Use the trapezium rule with six ordinates to calculate an approximate value of $\int_0^5 \sqrt{25 - x^2}\,dx$. Check your answer by integration.

   (Ans. 19 sq. units 19.6 sq. units)

6. Obtain an approximate value to three decimal places of
   $$\int_0^{\frac{\pi}{2}} \sqrt{\cos \theta}\,d\theta.$$
   (i) Using the trapezoidal rule with 6 intervals
   (ii) Using Simpson's Rule with 7 ordinates.

   (Ans. (i) 1.17 sq. units   (ii) 1.188 sq. units).

7. Use
   (i) The trapezium rule with 6 ordinates
   (ii) Simpson's Rule with 4 intervals to evaluate approximately
   $$\int_1^2 \sqrt{1 + x^2}\,dx.$$
   (Ans. (i) 1.812 sq. units   (ii) 1.810 sq. units.)

8. Use Simpson's Rule with 5 ordinates to evaluate approximately $\int_1^3 \sqrt{\ln x}\, dx$.

(Ans. 1.515 sq. units)

9. Use Simpson's Rule with 5 ordinates to evaluate approximately the following

(i) $\int_0^{\frac{1}{2}} \sqrt{1-x^2}\, dx$

(ii) $\int_0^1 \frac{1}{\sqrt{1+x^2}}\, dx$

(iii) $\int_0^2 e^{-x}\, dx$

(iv) $\int_1^4 \log x\, dx$

(v) $\int_0^{\frac{\pi}{6}} \sqrt{\sec x}\, dx$.

(Ans. (i) 0.478 (ii) 0.881 (iii) 0.865 (iv) 1.105 (v) 0.536 square units.)

10. Use Simpson's Rule, with 7 ordinates (or 6 intervals), to find an approximate value for $\int_0^\pi \sqrt{1+\cos x}\, dx$, giving your answer to three significant figures. (Ans. 2.828 sq. units)

11. Use the trapezium rule with 6 intervals to find an approximation to

$\int_0^6 \frac{1}{\sqrt{1+x^3}}\, dx$. (Ans. 1.974 sq. units)

12. Determine the definite integral $\int_{-2}^2 \frac{e^x}{1+e^x}\, dx$.

(i) Using the trapezoidal rule with five ordinates.

(ii) Using the Simpson's Rule with five ordinates.

(iii) Using an appropriate substitution.

(Ans. (i) 2 (ii) 2 (iii) 2 sq. units.)

13. Find the area under the curve $y = 3x^2 + 2x + 1$ between $x = -3$ and $x = 2$ using

(i) integration

(ii) the trapezoidal rule with five intervals

(iii) Simpson's Rule with 11 ordinates

(iv) the mid-ordinate rule with 10 intervals.

Ans. (i) 35 sq.units (ii) 37.5 sq. units (iii) 35 sq. units (iv) 35.625 sq. units

14. Evaluate approximately the area $\int_0^1 \frac{dx}{(1+x^2)^{\frac{1}{2}}}$ using

(i) Simpson's Rule with 4 intervals

(ii) Trapezoidal rule with 6 ordinates.

Comment about the number of ordinates and intervals in using these two rules.

(Ans. (i) 0.881 sq. units (ii) 0.880 sq. units).

15. Find approximately the definite integral $\int_1^2 10^x\, dx$ using Simpson's Rule with eleven ordinates.

(Ans. 39.1 sq. units.)

16. If the radius of a circle is 14 cm, determine the area of the circle by considering 15 ordinates when the centre is at the origin, for the area of the quarter of the circle and then multiplying the result by 4. Use Simpson's Rule.

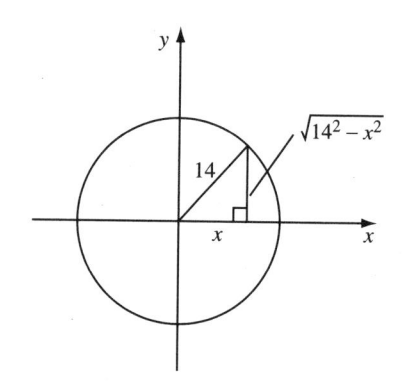

Fig. 7-I/31

Check this approximate area with the formula πr^2. What is the percentage error?

Determine $\int_0^{14} \sqrt{14^2 - x^2}\, dx$ exactly.

(Ans. 153.5 cm², 614 cm², 615.8 cm², 0.3%, 153.94 cm²)

17. Find the approximate values of the integral $\int_0^\pi \sqrt{\sin x}\, dx$ using 10 intervals and employing

(i) Trapezoidal rule

(ii) Simpson's Rule.

(Ans. (i) 2.323 sq. units, (ii) 2.368 sq. units)

18. Find the approximate values of the integral $\int_2^5 \ln x\, dx$ for 5 ordinates using Simpson's Rule.

(Ans. 3.661 sq. units.)

10

Determines the Volume of Revolution of a Simple Area and Locates the Position of its Centroid

Sketches a typical incremental volume of a simple function.

Determines the volume of the increment in a suitable notation

Let $f(x)$ be a curve as shown

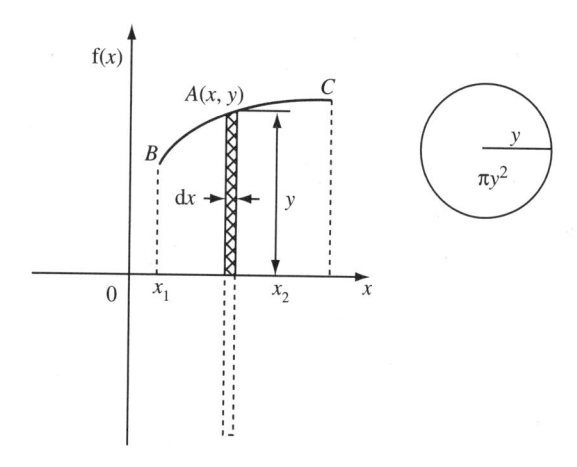

Fig. 7-I/32

Consider an elemental strip of width dx, the coordinates of a point A on the curve are (x, y), the height of the elemental rectangle is y. Revolving this strip about the x-axis, an elemental circular disc is obtained of an incremental volume $\pi y^2 \, dx$.

States the limits of the variable defining the required volume.

Revolving the arc BAC about the x-axis the volume so obtained lies between the limits $x = x_1$, and $x = x_2$.

The summation of all these elemental volumes $\pi y^2 \, dx$ lies between $x = x_1$ and $x = x_2$, this is represented by the integral.

$$\int_{x=x_1}^{x=x_2} \pi y^2 \, dx = \pi \int_{x=x_1}^{x=x_2} [f(x)]^2 \, dx$$

where $f(x) = y$.

Determines the required volume by summing up, the incremental volumes using definite integration.

WORKED EXAMPLE 84

Determine the volume generated by revolving the arc $y = \sqrt{x}$ about the x-axis between the limits $x = 1$ and $x = 3$.

Solution 84

$$V = \int_1^3 \pi y^2 \, dx$$

$$= \pi \int_1^3 \left(\sqrt{x}\right)^2 \, dx = \pi \left[\frac{x^2}{2}\right]_1^3 = 4\pi \text{ cubic units.}$$

WORKED EXAMPLE 85

Determine also the volume generated by revolving the arc $y = \sqrt{x}$ about the y-axis between the limits $y = 1$ and $y = 2$.

Solution 85

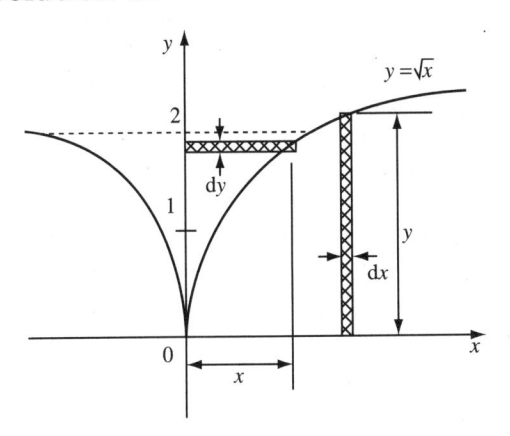

Fig. 7-I/33

$$V = \pi \int_1^2 x^2 \, dy = \pi \int_1^2 y^4 \, dy = \pi \left[\frac{y^5}{5}\right]_1^2$$

$$= \pi \left[\frac{32}{5} - \frac{1}{5}\right] = 31\frac{\pi}{5} \text{ cubic units}$$

$y = \sqrt{x}, y^2 = x, y^4 = x^2.$

WORKED EXAMPLE 86

Sketch the line $y = 3x$ between the values of $x = 0$ and $x = 3$. A cone is now formed by the revolution of the line about the x-axis determine the volume of the cone.

Solution 86

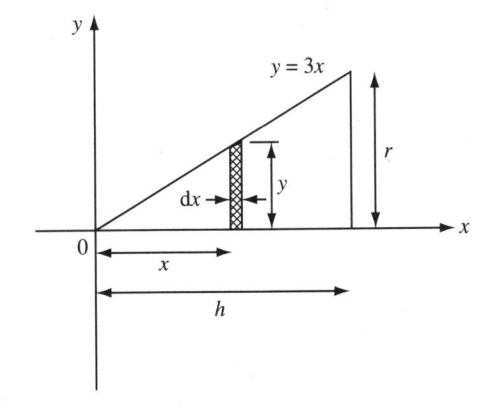

Fig. 7-I/34

Consider an elemental strip under the line as shown.

The revolution of the elemental area $y \, dx$ about the x-axis generates a circular disc of width dx.

$dv = \pi y^2 \, dx$ the elemental volume summing up all these strips between $x = 0$ and $x = 3$, we have

$$\int_0^h dv = \int_0^h \pi y^2 \, dx.$$

Let r be the radius of the cone and h the height. From the similar triangles

$$\frac{y}{x} = \frac{r}{h}$$

$$y = \frac{r}{h} x$$

$$V = \int_0^h \pi \left(\frac{r^2}{h^2} x^2\right) dx = \int_0^h \frac{\pi r^2}{h^2} x^2 \, dx$$

$$= \frac{\pi r^2}{h^2} \left[\frac{x^3}{3}\right]_0^h$$

$$\boxed{V = \frac{1}{3}\pi r^2 h}$$ the volume of a right circular cone of height h and radius r.

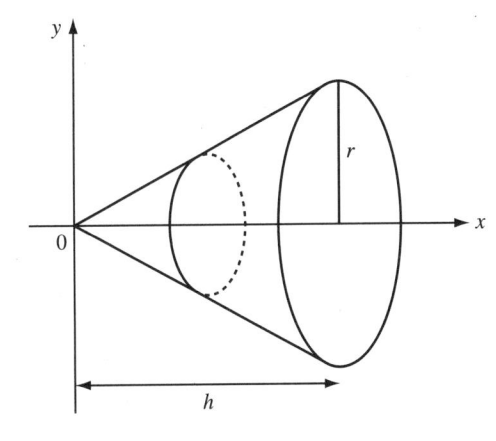

Fig. 7-I/35

WORKED EXAMPLE 87

Prove that the volume of a sphere of radius a is
$$V = \frac{4}{3}\pi a^3$$

Solution 87

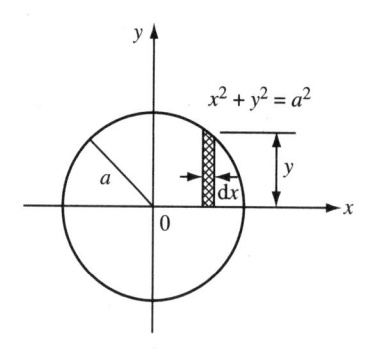

Fig. 7-I/36

If we revolve a circle about the x-axis or y-axis, a sphere is generated. The equation of a circle with its centre at the origin is $x^2 + y^2 = a^2$.

Consider a strip of height y and width dx. The strip when is revolved through one complete revolution about the x-axis generated an elemental volume $dv = \pi y^2\,dx$.

The total volume of the sphere so generated

$$V = \int_{-a}^{a} \pi y^2\,dx = \int_{-a}^{a} \pi\left(a^2 - x^2\right) dx = \left[\pi a^2 x - \pi \frac{x^3}{3}\right]_{-a}^{a}$$

$$V = \pi a^2 a - \pi \frac{a^3}{3} - \pi a^2(-a) + \frac{\pi(-a)^3}{3}$$

$$V = \pi a^3 - \frac{\pi a^3}{3} + \pi a^3 - \frac{\pi a^3}{3}$$

$$= 2\pi a^3 - \frac{2\pi a^3}{3} = \frac{4\pi a^3}{3}$$

$$V = \frac{4}{3}\pi a^3 \text{ is the volume of the sphere.}$$

Determines the position of its centroid by integration

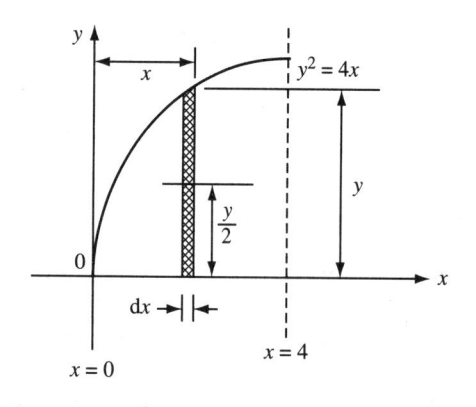

Fig. 7-I/37

Part of the parabola $y^2 = 4x$ is shown in the diagram, determine the centre of gravity of the area under the curve between the limit $x = 0$ and $x = 4$. An elemental area is shown when the height is y and the width dx, the first moment of the element about the x-axis is $(y\,dx) \times \frac{1}{2}y = $ the elemental area $\times$ distance of the centre of the elemental area which $\frac{1}{2}y$.

The total first moment about the x-axis is given by the integral

$$\int_{x=0}^{x=4} (y\,dx)\frac{1}{2}y = \frac{1}{2}\int_0^4 y^2\,dx = \frac{1}{2}\int_0^4 4x\,dx = 2\int_0^4 x\,dx$$

$$= \left[x^2\right]_0^4 = 16.$$

The area under the curve is found

$$\int_0^4 y\,dx = \int_0^4 \sqrt{4x}\,dx = \int_0^4 2x^{\frac{1}{2}}\,dx$$

$$= 2\left[\frac{x^{\frac{3}{2}}}{\frac{3}{2}}\right]_0^4 = \frac{4}{3}\left[4^{\frac{3}{2}}\right] = \frac{4}{3}(2)^3 = \frac{32}{3}.$$

To find $\bar{x} = $ the x-coordinate of the centre of gravity is found

$$\bar{x} = \frac{\text{First moment about the } x\text{-axis}}{\text{the area under the curve}}$$

$$\bar{x} = \frac{16}{\frac{32}{3}} = \frac{3}{2}$$

To find $\bar{y} = $ the y-coordinate of the centre of gravity, we take moment about the y-axis.

First moment about the y-axis

$$= \int_0^4 (y\,dx)x = \int_0^4 x\sqrt{4x}\,dx = \int_0^4 2x^{\frac{3}{2}}\,dx$$

$$= 2\left[\frac{x^{\frac{5}{2}}}{\frac{5}{2}}\right]_0^4 = \frac{4}{5}\left[4^{\frac{5}{2}}\right] = \frac{4}{5} \times (2^2)^{\frac{5}{2}} = \frac{128}{5}$$

$$\bar{y} = \frac{\text{First moment about the } y\text{-axis}}{\text{the area under the curve}}$$

$$= \frac{\frac{128}{5}}{\frac{32}{3}} = \frac{12}{5} = 2.4$$

Determines the Volume of Revolution of a Simple Area and Locates the Position of its Centroid — 75

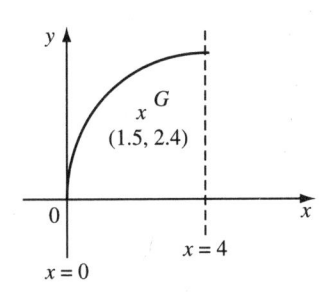

Fig. 7-I/38

The coordinates of the centre of gravity are (1.5, 2.4).

WORKED EXAMPLE 88

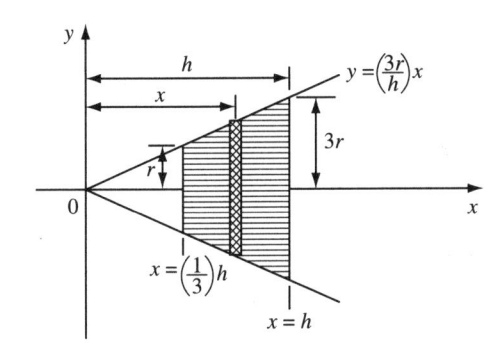

Fig. 7-I/39

Find the coordinates of the centre of gravity of the frustum of the cone whose dimensions are shown.

Solution 88

The elemental volume of the elemental strip of height $2y$ and width dx is $\pi y^2 \, dx$, its moment about the y-axis is $x\pi y^2 \, dx$ and the total first moment

$$\text{First moment} = \int_{\frac{h}{3}}^{h} \pi x y^2 \, dx = \int_{\frac{h}{3}}^{h} \pi x \left(\frac{3r}{h} x\right)^2 dx$$

$$= \pi \frac{9r^2}{h^2} \left[\frac{x^4}{4}\right]_{\frac{h}{3}}^{h}$$

$$= \frac{\pi 9 r^2}{h^2} \left[\frac{h^4}{4} - \frac{h^4}{81 \times 4}\right]$$

$$= \frac{9}{4} \frac{\pi r^2}{h^2} h^4 \frac{80}{81} = \frac{20}{9} \pi r^2 h^2.$$

The volume $= \int_{\frac{h}{3}}^{h} \pi y^2 \, dx = \pi \frac{9r^2}{h^2} \int_{\frac{h}{3}}^{h} x^2 \, dx$

$$= \frac{\pi 9 r^2}{h^2} \left[\frac{x^3}{3}\right]_{\frac{h}{3}}^{h}$$

$$= \frac{9\pi r^2}{h^2} h^3 \left[\frac{1}{3} - \frac{1}{81}\right]$$

$$= \pi r^2 h \frac{26}{9}.$$

$$\bar{x} = \frac{20}{9} \pi r^2 h^2 \times \frac{1 \times 9}{\pi r^2 h 26} = \frac{10}{13} h$$

$$\frac{10}{13} h - \frac{h}{3} = \frac{30h - 13h}{39}$$

$$= \frac{17}{39} h.$$

Fig. 7-I/40

∴ the coordinates are $\left(\frac{17}{39} h, 0\right)$.

Exercises 10

1. Determine the volume obtained by rotating the area under between the curve $y^2 = 8x$, the x-axis and between $x = 1$ and $x = 3$ about the axis of x.

2. Determine the volume obtained by rotating the area under between the curve $x^2 = 4y$, the y-axis and between $y = 0$ and $y = 2$ about the y-axis.

3. Determine the volume obtained by rotating the area between the curves $x^2 = y$ and $y^2 = x$ about the x-axis.

4. Determine the volume obtained by rotating the area of the line $y = 3x + 5$ about the x-axis between $x = -1$ and $x = 5$.

5. Determine the volume of an ellipsoid by rotating the ellipse $\dfrac{x^2}{9} + \dfrac{y^2}{16} = 1$ about the x-axis.

6. Determine the volume generated by rotating the curve $y = \dfrac{1}{x}$ between $x = 2$ and $x = 4$ about the x-axis.

7. Determine the volume generated by rotating the curve $y = e^{2x}$ between $x = 0$ and $x = 3$ about the x-axis.

8. Determine the volume generated by rotating the curve $y = 3\sin x$ between $x = 0$ and $x = \pi$ about the x-axis.

9. Determine the coordinates of the centroid of the curve $\omega y = \sin x$ between ($x = 0$ and $x = \frac{\pi}{2}$)

10. Determine the coordinates of the centroid of the area $y = 1 - x^2$ and the x-axis.

11. Find the centre of gravity of the area between the curve $y^2 = x$, and the line $x = 2$.

12. Find the centre of gravity of the area between the curve $y^2 = 2x$ and the line $x = \dfrac{1}{2}$.

13. Find the centre of gravity of the area between the curve $x^2 = 4y$ and the line $y = 1$.

14. Find the coordinates of the centroid of the area enclosed by the curve $y^2 = x$ and the line $x = 5$.

15. Find the centre of gravity of the area enclosed between the curve $y = 3x^3$ the x-axis and $x = 3$.

11

Differential Equations Separable Variables

Solve the differential equations.

1. $\dfrac{dy}{dx} = x + 5$

2. $\dfrac{dy}{dx} = 3x^2 + 1$

3. $\dfrac{dy}{dx} = 2x^3 - 3x^2 + x - 1$

4. $\dfrac{dy}{dx} = \dfrac{1}{x}$

5. $\dfrac{dy}{dx} = \dfrac{1}{x^2} + \dfrac{1}{x} + 3$

6. $\dfrac{dy}{dx} = \dfrac{y}{x}$

7. $\dfrac{dy}{dx} = \dfrac{x}{y}$

8. $\dfrac{dy}{dx} = \dfrac{x^2}{y^2}$

9. $\dfrac{dy}{dx} = \dfrac{y^2}{x^2}$

10. $\dfrac{dy}{dx} = \left(\dfrac{y}{x}\right)^{\frac{1}{2}}$

11. $\dfrac{dy}{dx} = 5xy$

12. $\dfrac{dy}{dx} = 5x^2 y^2$

13. $\dfrac{dy}{dx} = (xy)^{\frac{1}{3}}$

14. $\dfrac{dz}{dx} = zx$

15. $\dfrac{dz}{dx} = z^2 x^2$

16. $\dfrac{dz}{dx} = \dfrac{\sin ax}{\cos ax}$

17. $\dfrac{dz}{dx} = \dfrac{\cos kx}{\sin kz}$

18. $\dfrac{dz}{dx} = \dfrac{\sin kx}{\cos kx}$

19. $\dfrac{dz}{dx} = \sin\left(x + \dfrac{\pi}{4}\right)$

20. $\dfrac{dx}{dt} = e^{-t}$

21. $\dfrac{dx}{dt} = e^{3x}$

22. $\dfrac{dx}{dt} = \dfrac{e^{-x^2}}{x}$

23. $\dfrac{dy}{dt} = t^3 (1 + 4t^4)^{-\frac{5}{3}}$

24. $\dfrac{dy}{dt} = (3t - t^{-3})^3 (3 + 3t^{-4})$

   $t = 1$, when $y = 2$

77

25. $\dfrac{du}{dx} = 7x^2 + 3x - 1$

　　$u = 1$, when $x = 0$

26. $\dfrac{du}{dx} = x^{\frac{1}{3}} + x^{\frac{1}{5}}$

　　$x = 0$, when $u = -1$

27. $\dfrac{dx}{dt} = \dfrac{\sin^5 2x}{\cos 2x}$

　　$t = 1$, when $x = \dfrac{\pi}{4}$

28. $\dfrac{dy}{dx} = \dfrac{\ln x}{x}$

29. $\dfrac{dy}{dx} = \dfrac{y}{\ln y}$

30. $\dfrac{dy}{dx} = \dfrac{e^{3x}}{e^{3y}}$

31. $\dfrac{dy}{dx} = \dfrac{\sin 3x}{\cos 3y}$

32. $\dfrac{dy}{dx} = \dfrac{\cos 5y}{\sin 5x}$

33. $\dfrac{dy}{dx} = \dfrac{e^{-x}}{e^{-3y}}$

34. $\dfrac{dy}{dx} = e^{5x-7y}$

35. $\dfrac{dy}{dx} = e^{x+y+2}$

36. $x\dfrac{dy}{dx} = 5y$

37. $x(y+2)\dfrac{dy}{dx} = y^2$

38. $x^4 \, dx + (y+3)^3 \, dy = 0$

39. $(1 - \cos\theta) \, dx - 3e^{-3x} \, d\theta = 0$

　　$\theta = \dfrac{\pi}{2}, x = 0$

40. $x^3(y-1) \, dx - y^4(x+1) \, dy = 0$

41. $(1 + x^4) \, dy - x^3 y \, dx = 0$

　　$y = 3$ when $x = 1$.

12

Exact Differential Equations

Find the general solution of each of the following exact differential equations.

1. $-\dfrac{x}{y^2}\dfrac{dy}{dx} + \dfrac{1}{y} = x^2$

2. $\dfrac{1}{x}\dfrac{dy}{dx} - \dfrac{y}{x^2} = 2x + 1$

3. $e^y x \dfrac{dy}{dx} + e^y = \sin x$

4. $e^x e^y + e^x e^y \dfrac{dy}{dx} = \cos x$

5. $-\sin x \, \sin y \dfrac{dy}{dx} + \cos x \cos y = e^x$

6. $-e^x \sin y \dfrac{dy}{dx} + e^x \cos y = e^{-2x}$

7. $e^y \sin x \dfrac{dy}{dx} + e^y \cos x = e^{\frac{1}{3}x}$

8. $e^y \tan x \dfrac{dy}{dx} + \sec^2 x \, e^y = x \ln x$

9. $\sec x \, e^y \dfrac{dy}{dx} + \sec x \tan x \, e^y = \ln x$

10. $-\operatorname{cosec} x \cot x \, e^y + \operatorname{cosec} x \, e^y \dfrac{dy}{dx} = e^x$

11. $\operatorname{cosec} x \dfrac{dy}{dx} - y \operatorname{cosec} x \cot x = e^{-3x}$

12. $\cot x \dfrac{dy}{dx} - y \operatorname{cosec}^2 x = 5$

13. $\dfrac{x}{y}\dfrac{dy}{dx} + \ln y = e^x$

14. $x e^y \dfrac{dy}{dx} + e^y = x^2$

15. $x^2 \cos y \dfrac{dy}{dx} + 2x \sin y = \cos x$

16. $x \sec^2 y \dfrac{dy}{dx} + \tan y = x^2 + 3x + 1.$

13

The Integrating Factor

$$\frac{dy}{dx} + Py = Q \qquad \ldots (1)$$

is a first order differential equation where P and Q are functions of x. Before embarking in finding a solution for (1) consider the product rule

$$\frac{d}{dx}(uv) = \left(\frac{du}{dx}\right)v + u\left(\frac{dv}{dx}\right) \text{ or}$$

$$d(uv) = (du)v + u(dv) \text{ or}$$

$$(du)v + u(dv) = d(uv) \text{ and as an example}$$

$$\frac{d(x \sin x)}{dx} = \sin x + x \cos x \text{ or}$$

$$\frac{d}{dx}(x \tan y) = \tan y + x \sec^2 y \frac{dy}{dx}$$

$\tan y + x \sec^2 y \dfrac{dy}{dx}$ is to be recognised as the derivative of the product $x \tan y$ with respect to x.

A differential equation such as

$$\tan y + x \sec^2 y \frac{dy}{dx} = \cos x$$

can be written as

$$\frac{d}{dx}(x \tan y) = \cos x$$

$$d(x \tan y) = \cos x \, dx \qquad \ldots (2)$$

Integrating both sides of equation (2)

$$\int d(x \tan y) = \int \cos x \, dx$$

the solution is $x \tan y = \sin x + c$ and it is called an exact differential equation.

If, however, the equation is not an exact one then we should multiply both sides of equation (1) by a factor called the integrating factor, in order to make the left hand side an exact differential equation.

Let the integrating factor, I.F. be equal to $e^{\int P dx}$, which is to be proved in the following

$$(\text{I.F.})\frac{dy}{dx} + (\text{I.F.})Py = (\text{I.F.})Q.$$

The left hand side (L.H.S.) of this equation is to be equal to $u\dfrac{dv}{dx} + v\dfrac{du}{dx}$

$$(\text{I.F.})\frac{dy}{dx} + (\text{I.F.})Py = u\frac{dv}{dx} + v\frac{du}{dx}$$

$$v = \text{I.F.}, \frac{du}{dx} = \frac{dy}{dx}, u = y \text{ and } \frac{dv}{dx} = (\text{I.F.})P$$

$$\frac{dv}{dx} = \frac{d(\text{I.F.})}{dx}$$

$$d(\text{I.F.}) = (\text{I.F.})P$$

$$\frac{d(\text{I.F.})}{\text{I.F.}} = P.$$

Integrating both sides with respect to x

$$\int \frac{d(\text{I.F.})}{\text{I.F.}} dx = \int P \, dx$$

$$\ln(\text{I.F.}) = \int P \, dx$$

by the definition of the logarithm $\boxed{e^{\int P dx} = \text{I.F.}}$

Therefore, this equation is an integrating factor (I.F.) for the equation $\dfrac{dy}{dx} + Py = Q$.

WORKED EXAMPLE 89

Solve the differential equation
$$\frac{dy}{dx} + y \tan x = \cos x$$
using the integrating factor method. If $x = \frac{\pi}{4}, y = 1$, find y in terms of x.

Solution 89

$$\frac{dy}{dx} + y \tan x = \cos x \qquad \ldots (1)$$

This equation is of the form $\frac{dy}{dx} + Py = Q$ where P and Q are functions of x,

$P = \tan x$ and $Q = \cos x$.
The integrating factor

$$\text{I.F.} = e^{\int \tan x \, dx}$$
$$= e^{\int \frac{\sin x}{\cos x} \, dx} = e^{-\int \frac{d(\cos x)}{\cos x}}$$
$$= e^{-\ln \cos x} = e^{\ln \sec x} = \sec x.$$

Multiplying each side of the equation (1) with the integrating factor

$$\sec x \frac{dy}{dx} + y \tan x \sec x = \cos x \sec x \text{ the derivative of the product } y \cdot \sec x$$

$$\frac{d}{dx}(y \cdot \sec x) = \cos x \sec x = 1$$

$$d(y \sec x) = dx.$$

Integrating both sides with respect to x

$$\int d(y \sec x) = \int dx$$

$$y \sec x = x + c$$

$$x = \frac{\pi}{4}, y = 1$$

$$1 \cdot \sec \frac{\pi}{4} = \frac{\pi}{4} + c$$

$$c = \sqrt{2} - \frac{\pi}{4}$$

$$y \sec x = x + \sqrt{2} - \frac{\pi}{4}$$

Therefore $\boxed{y = \left(x + \sqrt{2} - \frac{\pi}{4}\right) \cos x.}$

WORKED EXAMPLE 90

Solve the differential equation
$$x \frac{dy}{dx} + y = xe^{-x} \quad \text{if } x = 0 \text{ when } y = 0.$$

Solution 90

$$x \frac{dy}{dx} + y = xe^{-x}.$$

This can be written in the form $\frac{dy}{dx} + Py = Q$ where P and Q are functions of x.

$$\frac{dy}{dx} + \frac{1}{x} y = e^{-x} \qquad \ldots (1)$$

where $P = \frac{1}{x}$ and $Q = e^{-x}$.

The I.F. $= e^{\int P dx} = e^{\int \frac{1}{x} dx} = e^{\ln x} = x.$

To show that $e^{\ln x} = x$, take logs on both sides to the base e.

$$\ln e^{\ln x} = \ln x$$

$$\ln x \ln e = \ln x.$$

This L.H.S. is $\ln x$ since $\ln e = 1$.
Multiplying each term of (1) by x.

$$x \frac{dy}{dx} + y = x \, e^{-x}$$

$$\frac{d}{dx}(y \cdot x) = x \, e^{-x}$$

$$d(y \cdot x) = x \, e^{-x} dx.$$

Integrating both sides with respect to x.

$$\int d(yx) = \int x \, e^{-x} dx$$

$$yx = (-e^{-x})x - \int -e^{-x} \cdot 1 \, dx$$

$$= -xe^{-x} + \int e^{-x} dx = -xe^{-x} - e^{-x} + c.$$

If $x = y = 0$,

$$0 = -e^0 + c \quad \therefore c = 1$$

$$yx = -xe^{-x} - e^{-x} + 1$$

$$\therefore \boxed{y = -e^{-x} - \frac{1}{x} e^{-x} + \frac{1}{x}}$$

Worked Example 91

Find the general solution
$(1+x^2)\dfrac{dy}{dx} = x(1+x+y)$. If $x=0$, when $y=1$.

Solution 91

$(1+x^2)\dfrac{dy}{dx} = x(1+x+y)$. This equation may be written in the form

$$\dfrac{dy}{dx} + Py = Q$$

$$(1+x^2)\dfrac{dy}{dx} = x(1+x+y)$$

$$(1+x^2)\dfrac{dy}{dx} = x + x^2 + xy$$

$$(1+x^2)\dfrac{dy}{dx} - xy = x(x+1)$$

$$\dfrac{dy}{dx} - \dfrac{x}{1+x^2}y = \dfrac{x(x+1)}{1+x^2} \qquad \ldots (1)$$

where $P = -\dfrac{x}{1+x^2}$ and $Q = \dfrac{x(x+1)}{1+x^2}$.

The I.F. $= e^{\int P\,dx} = e^{\int -\frac{x}{1+x^2}\,dx}$

$$= e^{\frac{1}{2}\int -\frac{d(1+x^2)}{(1+x^2)}} = e^{-\frac{1}{2}\ln(1+x^2)}$$

$$= e^{\ln\frac{1}{(1+x^2)^{\frac{1}{2}}}}$$

$$\text{I.F.} = \dfrac{1}{\sqrt{1+x^2}}.$$

Multiplying each term of equation (1) by the I.F.

$$\dfrac{1}{\sqrt{1+x^2}} \cdot \dfrac{dy}{dx} - \dfrac{x}{(1+x^2)^{\frac{3}{2}}}y = \dfrac{x(x+1)}{(1+x^2)^{\frac{3}{2}}}$$

$$\dfrac{d}{dx}\left(y \cdot \dfrac{1}{\sqrt{1+x^2}}\right) = \dfrac{x(x+1)}{(1+x^2)^{\frac{3}{2}}}$$

$$d\left(y\dfrac{1}{\sqrt{1+x^2}}\right) = \dfrac{x(x+1)}{(1+x^2)^{\frac{3}{2}}}\,dx.$$

Integrating both sides with respect to x.

$$\int d\left(y\dfrac{1}{\sqrt{1+x^2}}\right) = \int \dfrac{x(x+1)}{(1+x^2)^{\frac{3}{2}}}\,dx.$$

$$y\dfrac{1}{\sqrt{1+x^2}} = \int \dfrac{x(x+1)}{(1+x^2)^{\frac{3}{2}}}\,dx.$$

let $x = \tan\theta \quad \dfrac{dx}{d\theta} = \sec^2\theta$

$$= \int \dfrac{\tan^2\theta + \tan\theta}{(\tan^2\theta + 1)^{\frac{3}{2}}}\sec^2\theta\,d\theta$$

$$= \int \dfrac{\tan\theta(1+\tan\theta)}{\sec^3\theta}\sec^2\theta\,d\theta$$

$$= \int \tan\theta(1+\tan\theta)\cos\theta\,d\theta$$

$$= \int \sin\theta(1+\tan\theta)\,d\theta$$

$$= \int \left(\sin\theta + \dfrac{\sin^2\theta}{\cos\theta}\right)d\theta$$

$$= \int \left(\sin\theta + \dfrac{1-\cos^2\theta}{\cos\theta}\right)d\theta$$

$$= \int (\sin\theta + \sec\theta - \cos\theta)\,d\theta$$

$$= -\cos\theta + \ln(\sec\theta + \tan\theta) - \sin\theta + c$$

$$y\dfrac{1}{\sqrt{1+x^2}} = -\dfrac{1}{\sqrt{1+x^2}} + \ln\left(\sqrt{1+x^2}+x\right)$$

$$\qquad - \dfrac{x}{\sqrt{1+x^2}} + c$$

$$1 \cdot \dfrac{1}{1} = -1 + \ln 1 - 0 + c$$

$$\boxed{c = 2}$$

$$\boxed{\begin{array}{l} y = -1 + \sqrt{1+x^2}\ln\left(\sqrt{1+x^2}+x\right) \\ \quad - x + 2\sqrt{1+x^2} \end{array}}$$

The Integrating Factor

WORKED EXAMPLE 92

Find the general solution

$$\frac{dx}{d\theta} + \cot\theta \, x = \cos\theta. \text{ If } \theta = \frac{\pi}{2} \text{ for } x = 1.$$

Solution 92

$$\frac{dx}{d\theta} + \cot\theta \, x = \cos\theta$$

$$\text{I.F.} = e^{\int \cot\theta \, d\theta} = e^{\ln\sin\theta} = \sin\theta$$

$$\sin\theta \frac{dx}{d\theta} + \cot\theta \, x \, \sin\theta = \cos\theta \sin\theta$$

$$\frac{d}{d\theta}(x \sin\theta) = \frac{\sin 2\theta}{2}$$

$$d(x \sin\theta) = \frac{1}{2} \sin 2\theta \, d\theta$$

$$\int d(x \sin\theta) = \frac{1}{2} \int \sin 2\theta \, d\theta$$

$$x \sin\theta = -\frac{\cos 2\theta}{4} + c$$

$$1 \cdot \sin\frac{\pi}{2} = -\frac{\cos 2\theta}{4} + c$$

$$c = 1 + \frac{\cos\pi}{4} = \frac{3}{4}$$

$$x \sin\theta = -\frac{\cos 2\theta}{4} + \frac{3}{4}$$

$$x = -\frac{\cos 2\theta}{4\sin\theta} + \frac{3}{4\sin\theta}$$

$$\boxed{x = \frac{3}{4}\operatorname{cosec}\theta - \frac{1}{4}\cos 2\theta \operatorname{cosec}\theta}$$

Exercises 13

1. $\sin x \dfrac{dy}{dx} + y \cos x = \sin 2x$

2. $\dfrac{dy}{dx} = \dfrac{1}{x} y + x^2 e^x$.

3. $\dfrac{dy}{dx} + \dfrac{2x}{1+x^2} y = \dfrac{x}{1+x^2}$.

4. $\dfrac{dy}{dx} - y \cot x = \sin x$.

5. $(1 - x^2)\dfrac{dy}{dx} - 2xy = 1 + x$.

6. $(1 - x^2)\dfrac{dy}{dx} + 2xy = 1 + x$.

7. $\dfrac{dx}{dt} - 3x = \sin 3t$.

8. $\dfrac{dy}{dx} + y \sec x \operatorname{cosec} x = \tan x$.

9. $\dfrac{dy}{dx} + y = x^2$.

10. $\dfrac{dy}{dx} - \dfrac{1}{1+x} y = x$.

11. $\dfrac{dy}{dx} - y \sec x \operatorname{cosec} x = \cot x$.

12. $\dfrac{dy}{dx} + \dfrac{\frac{1}{x}}{\ln x} y = x^3$.

13. $\dfrac{dy}{dx} + y \dfrac{x}{1+x} = \dfrac{1}{(1+x)e^{x+1}}$.

14. $\dfrac{dy}{dx} + y \cot x = \sin 2x$.

15. $\dfrac{dy}{dx} + \dfrac{2x}{1-x^2} y = \dfrac{x^3}{1-x^2}$.

16. $\dfrac{dy}{dx} - y \cot x = 5 \sin x \cos x$.

17. $\dfrac{dy}{dx} + \dfrac{2x}{1+x^2} y = \dfrac{x^3}{1+x^2}$.

18. $x^2 \dfrac{dy}{dx} + xy = 5$.

14

Second Order Differential Equations

$$a\frac{d^2y}{dx^2} + b\frac{dy}{dx} + cy = f(x)$$

where a, b and c are constant coefficients and $f(x)$ is a function of x.

If $f(x) = 0$, then

$$\boxed{a\frac{d^2y}{dx^2} + b\frac{dy}{dx} + cy = 0} \quad \cdots \text{(a)}$$

Let $y = u$ where u is a function of x

$$a\frac{d^2u}{dx^2} + b\frac{du}{dx} + cu = 0 \quad \cdots (1)$$

Let v be another function of x, let $y = v$, then

$$a\frac{d^2v}{dx^2} + b\frac{dv}{dx} + cv = 0 \quad \cdots (2)$$

Adding equations (1) and (2)

$$a\left(\frac{d^2u}{dx^2} + \frac{d^2v}{dx^2}\right) + b\left(\frac{du}{dx} + \frac{dv}{dx}\right)$$
$$+ c(u+v) = 0 \quad \cdots (3)$$

and since

$$\frac{d}{dx}(u+v) = \frac{du}{dx} + \frac{dv}{dx} \text{ and } \frac{d^2}{dx^2}(u+v) = \frac{d^2u}{dx^2} + \frac{d^2v}{dx^2}$$

then equation (3) can be written

$$\frac{a d^2}{dx^2}(u+v) + \frac{b d}{dx}(u+v) + c(u+v) = 0$$ which is the original equation with y replaced by $u+v$.

Therefore, if $y = u$ and $y = v$ are solutions of the equation $a\frac{d^2y}{dx^2} + b\frac{dy}{dx} + c(u+v) = 0$, so $y = u + v$ is also a solution.

Let $y = Ae^{mx}$ be a solution of $a\frac{d^2y}{dx^2} + b\frac{dy}{dx} + cy = 0$

$$\frac{dy}{dx} = Ame^{mx}$$

$$\frac{d^2y}{dx^2} = Am^2e^{mx}.$$

(a) Substituting $y = Ae^{mx}$, $\frac{dy}{dx} = Ame^{mx}$ and $\frac{d^2y}{dx^2} = Am^2e^{mx}$ in equation A, $aAm^2e^{mx} + bAme^{mx} + cAe^{mx} = 0$, which is reduced to the equation

$$\boxed{am^2 + bm + c = 0.}$$

This equation is a quadratic equation and is called the AUXILIARY EQUATION. Let m_1 and m_2 be the two roots of this auxiliary equation and the two solutions of the equation (a) are $\boxed{y = Ae^{m_1x}}$ and $\boxed{y = Be^{m_2x}.}$

Then the solution of equation (a) is

$$\boxed{y = Ae^{m_1x} + Be^{m_2x}}$$

where A and B are arbitrary constants.

The auxiliary equation of $\frac{d^2y}{dx^2} + 2\frac{dy}{dx} + 3y = 0$ is $m^2 + 2m + 3 = 0$.

If the auxiliary equation has factors $(m-1)(m+2) = 0$, $m^2 + m - 2 = 0$ then the second order differential equation is $\frac{d^2y}{dx^2} + \frac{dy}{dx} - 2y = 0$ and its solution is

$$y = Ae^x + Be^{-2x}.$$

The quadratic auxiliary equation, however, may have the following conditions:

(i) real and different roots
(ii) real and equal roots
(iii) complex roots

depending on the value of the discriminant, $D = b^2 - 4ac$. If D is positive, then the roots are real and different, if D is equal to zero, then the roots are real and equal and finally if D is negative, then the roots are complex.

Equal and Different Roots

WORKED EXAMPLE 93

Solve the differential equation $\dfrac{d^2y}{dx^2} - 3\dfrac{dy}{dx} + 2y = 0$.

Solution 93

The auxiliary equation is $m^2 - 3m + 2 = 0$ or $(m-1)(m-2) = 0$ and the solution is therefore given as

$$y = Ae^x + Be^{2x}$$

WORKED EXAMPLE 94

Solve the differential equation $\dfrac{d^2y}{dx^2} - \dfrac{dy}{dx} - 6y = 0$.

Solution 94

The auxiliary equation is $m^2 - m - 6 = 0$ which factorises to $(m+2)(m-3) = 0$ and hence $m_1 = -2$ and $m_2 = 3$. The solution is $y = Ae^{-2x} + Be^{3x}$.

Real and Equal Roots

Solve the differential equation

$$\dfrac{d^2y}{dx^2} - 2\dfrac{dy}{dx} + y = 0 \qquad \ldots (1)$$

which has an auxiliary equation

$$m^2 - 2m + 1 = 0$$

which factorises to $(m-1)(m-1) = 0$ $m = 1$ (twice).

If $m_1 = 1$ and $m_2 = 1$ then

$y = Ae^x + Be^x$ or $y = (A+B)e^x = Ce^x$

which implies that there is one arbitrary constant but we showed earlier that every second order differential equation has two arbitrary constants. If $y = Mxe^x$, then $\dfrac{dy}{dx} = Me^x + Mxe^x$ and $\dfrac{d^2y}{dx^2} = Me^x + Me^x + Mxe^x$ or $\dfrac{d^2y}{dx^2} = 2Me^x + Mxe^x$, substituting these values in (1) above, we have $2Me^x + Mxe^x - 2(Me^x + Mxe^x) + Mxe^x = 0$ the L.H.S. is zero and the solution $y = Mxe^x$ satisfies the equation.

The complete general solution is $y = Ae^x + Bxe^x$

$$\boxed{y = e^x(A + Bx)}$$

Therefore, a second order differential equation $a\dfrac{d^2y}{dx^2} + b\dfrac{dy}{dx} + cy = 0$ which has an auxiliary equation with real and equal roots then the general solution will be $\boxed{y = e^x(A + Bx)}$ where m is double root or in the above example $m = 1$ (twice).

WORKED EXAMPLE 95

Solve $\dfrac{d^2y}{dx^2} - 4\dfrac{dy}{dx} + 4y = 0$

Solution 95

The auxiliary equation for the above differential equation is $m^2 - 4m + 4 = 0$ or $(m-2)^2 = 0$ or $m = 2$ (twice). The general solution

$$\boxed{y = e^{2x}(A + Bx)}$$

Complex Roots

Let us suppose that the roots are complex $m_1 = 3 + i4$ and $m_2 = 3 - i4$ the auxiliary equation would be

$[m - (3 + i4)]$
$[m - (3 - i4)] = (m - 3 - i4)(m - 3 + i4)$
$\qquad = (m-3)^2 - i^2 16$
$\qquad = (m-3)^2 + 16$ where $i^2 = -1$
$\qquad = m^2 - 6m + 9 + 16$
$\qquad = m^2 - 6m + 25$.

Finally the second order differential equation will be $\dfrac{d^2y}{dx^2} - 6\dfrac{dy}{dx} + 25y = 0$.

The general solution will be
$$y = k_1 e^{(3+i4)x} + k_2 e^{(3-i4)x}$$
$$y = k_1 e^{3x} e^{i4x} + k_2 e^{3x} e^{-i4x}$$
where k_1, k_2 are arbitrary constants
$$y = e^{3x}\left(k_1 e^{i4x} + k_2 e^{-i4x}\right)$$
and from the knowledge of complex numbers
$e^{i4x} = \cos 4x + i\sin 4x$ and $e^{-i4x} = \cos 4x - i\sin 4x$
$$y = e^{3x}[k_1(\cos 4x + i\sin 4x) + k_2(\cos 4x - i\sin 4x)]$$
$$= e^{3x}[(k_1 + k_2)\cos 4x + i(k_1 - k_2)\sin 4x]$$
$$= e^{3x}[A\cos 4x + B\sin 4x]$$
where $A = k_1 + k_2$ and $B = i(k_1 - k_2)$ are the new arbitrary constants.

Therefore, if $m = 3 \pm i4$, the general solution can be written
$$y = e^{3x}[A\cos 4x + B\sin 4x] \text{ where } m = a \pm ib, a = 3 \text{ and } b = 4.$$

Therefore in general if the second order differential equation which has an auxiliary equation with complex roots $a + ib$ and $a - ib$, the general solution will be

$$\boxed{y = e^{ax}[A\cos bx + B\sin bx]}$$

a is the real part of the complex root and b is the imaginary part of the complex root. All the above work can now be summarised and should be remembered.

If $a\dfrac{d^2y}{dx^2} + b\dfrac{dy}{dx} + cy = 0$ then the general solutions will be

If $D > 0$ $\boxed{y = Ae^{m_1 x} + Be^{m_2 x}}$

If $D = 0$ $\boxed{y = e^{mx}(A + Bx)}$

If $D < 0$ $\boxed{y = e^{ax}(A\cos bx + B\sin bx)}$

where $m = a + ib$ is the complex root.

Special Cases of the Second Order Differential Equations

If $a = 0$ in $a\dfrac{d^2y}{dx^2} + b\dfrac{dy}{dx} + cy = 0$ then $b\dfrac{dy}{dx} + cy = 0$ this is now a first order differential equation of the separable variable form.

$$\frac{b\,dy}{dx} = -cy \qquad \frac{dy}{y} = -\frac{c}{b}dx.$$

Integrating both sides
$$\int \frac{dy}{y} = -\int \frac{c}{b}dx$$
$$\ln y = -\frac{c}{b}x + \text{constant}$$
by the definition of the logarithms
$y = e^{-\frac{c}{b}x+\text{constant}} = e^{-Ax}e^{B}$ and $y = A_1 e^{-Ax}$ or
$\boxed{y = Ae^{-mx}}$ where A is the arbitrary constant and $m = \dfrac{c}{b}$.

WORKED EXAMPLE 96

Solve $3\dfrac{dy}{dx} - 2y = 0$.

Solution 96

$y = Ae^{\left(\frac{2}{3}\right)x}$ where A is the arbitrary constant.

If $b = 0$ in $a\dfrac{d^2y}{dx^2} + b\dfrac{dy}{dx} + cy = 0$ then
$$a\frac{d^2y}{dx^2} + cy = 0$$

this can be written as $\dfrac{d^2y}{dx^2} + ky = 0$ or as $\dfrac{d^2y}{dx^2} \pm n^2 y = 0$. The auxiliary equation is

$m^2 \pm n^2 = 0$

I $m^2 + n^2 = 0$, therefore $m = \pm\sqrt{-n^2} = \pm in$

II or $m^2 - n^2 = 0$, therefore $m = \pm n$.

Case I $m = a \pm ib, a = 0, b = n$ therefore $y = e^{0\cdot x}(A\cos nx + B\sin nx)$

or $\boxed{y = A\cos nx + B\sin nx}$ ...(1)

Case II $m_1 = n, m_2 = -n$

$$\boxed{y = Ae^{nx} + Be^{-nx}} \qquad \ldots(2)$$

but $e^{nx} = \cosh nx + \sinh nx$
         $e^{-nx} = \cosh nx - \sinh nx$.

Therefore, (2) can be written

$$\boxed{y = A_1 \cosh nx + B_1 \sinh nx}$$

All the above solutions for the three cases are called COMPLEMENTARY FUNCTIONS abbreviated as C.F., if $f(x)$, however, is not equal to nought (or zero), then we shall have an additional solution called the PARTICULAR INTEGRAL, abbreviated as P.I.

The complete general solution for the second order differential equation

$$a\frac{d^2y}{dx^2} + b\frac{dy}{dx} + cy = f(x)$$

$y = \text{C.F.} + \text{P.I.}$

$f(x)$ can have the following forms.

(I) $f(x) = a_1 + b_1 x + b_2 x^2$ (algebraic function)

(II) $f(x) = A \sin kx$ or $f(x) = B \sin kx$ or both (Circular functions)

(III) $f(x) = Ae^{kx}$ (exponential).

Let us consider an example for each of the above cases.

Worked Example 97

Solve $\dfrac{d^2y}{dx^2} + \dfrac{dy}{dx} - 12y = 3x + 5$.

Solution 97

There are two steps in the solution of the above

(I) $\dfrac{d^2y}{dx^2} + \dfrac{dy}{dx} - 12y = 0$

the auxiliary equation $m^2 + m - 12 = 0$ which factorises to $(m-3)(m+4) = 0$, $m_1 = 3$ and $m_2 = -4$.

The complementary function is $y = Ae^{3x} + Be^{-4x}$.

(II) The particular integral can be found as follows:

Let $y = ax + b$, a function of the same form as $y = 3x + 5$.

$$\frac{dy}{dx} = a$$

$$\frac{d^2y}{dx^2} = 0.$$

Substituting $\dfrac{d^2y}{dx^2} = 0$, $\dfrac{dy}{dx} = a$, $y = ax + b$ in the left hand side of the equation to be solved, we have

$$a - 12(ax + b) = 3x + 5$$

$$-12ax - 12b + a = 3x + 5$$

equating coefficients

$-12a = 3$ and $a = -\dfrac{1}{4}$

$a - 12b = 5$ and $-12b = 5 + \dfrac{1}{4} = \dfrac{21}{4}$

$$b = -\frac{21}{4 \times 12} = -\frac{7}{16}.$$

Therefore, the P.I. solution is given

$$y = -\frac{1}{4}x - \frac{7}{16}.$$

The general solution

$$y = \text{C.F.} + \text{P.I.} = Ae^{3x} + Be^{-4x} - \frac{1}{4}x - \frac{7}{16}$$

$$\boxed{y = Ae^{3x} + Be^{-4x} - \frac{1}{4}x - \frac{7}{16}}$$

Worked Example 98

Solve $6\dfrac{d^2y}{dx^2} - \dfrac{dy}{dx} - y = 10 \sin 2x$.

Solution 98

The auxiliary equation $6m^2 - m - 1 = 0$ which factorises to $(2m-1)(3m+1) = 0$,

$m_1 = \dfrac{1}{2}$ and $m_2 = -\dfrac{1}{3}$.

The C.F. solution

$$y = Ae^{\frac{1}{2}x} + Be^{-\frac{1}{3}x}.$$

The P.I. solution.

Let $y = a \sin 2x + b \cos 2x$

$$\frac{dy}{dx} = 2a \cos 2x - 2b \sin 2x$$

$$\frac{d^2y}{dx^2} = -4a \sin 2x - 4b \cos 2x$$

substituting these values in the d.e.

$6(-4a \sin 2x - 4b \cos 2x) - (2a \cos 2x - 2b \sin 2x)$
$- (a \sin 2x + b \cos 2x) = 10 \sin 2x$

$-24a \sin 2x - 24b \cos 2x - 2a \cos 2x + 2b \sin 2x$
$- a \sin 2x - b \cos 2x = 10 \sin 2x.$

Equating coefficients

$$-24a + 2b - a = 10$$

$$-25a + 2b = 10 \qquad \ldots (1)$$

$$-24b - 2a - b = 0$$
$$-2a - 25b = 0 \quad \ldots (2)$$
$$50a - 4b = -20$$
$$\underline{-50a - 625b = 0}$$
$$-629b = -20$$
$$\therefore b = \frac{20}{629}$$
$$a = -\frac{25}{2}b = -\frac{25}{2}\left(\frac{20}{629}\right) = -\frac{250}{629}.$$

The general solution
$$y = \text{C.F.} + \text{P.I.}$$

$$\boxed{y = Ae^{\frac{1}{2}x} + Be^{-\frac{1}{3}x} - \frac{250}{629}\sin 2x + \frac{20}{629}\cos 2x}$$

WORKED EXAMPLE 99

Solve $\dfrac{d^2y}{dx^2} - 2\dfrac{dy}{dx} + 5y = 5e^{2x}$.

Solution 99

The complementary function
$$\frac{d^2y}{dx^2} - 2\frac{dy}{dx} + 5y = 0$$
$$m = \frac{2 \pm \sqrt{4 - 20}}{2} = 1 \pm i2$$
$m_1 = 1 + i2$ and $m_2 = 1 - i2$
$a = 1$ and $b = 2$
the roots are complex $y = e^x(A\cos 2x + B\sin 2x)$.
The Particular Integral. Let $y = Ae^{2x}$
$$\frac{dy}{dx} = 2Ae^{2x}$$
$$\frac{d^2y}{dx^2} = 4Ae^{2x}.$$

Substituting these values
$$4Ae^{2x} - 2(2Ae^{2x}) + 5(Ae^{2x}) = 5e^{2x}$$
$$4Ae^{2x} - 4Ae^{2x} + 5Ae^{2x} = 5e^{2x}.$$

Equate coefficients

$$4A - 4A + 5A = 5$$
$$A = 1.$$

Therefore the P.I. solution $y = e^{2x}$.
The general solution
$$\boxed{y = e^x(A\cos 2x + B\sin 2x) + e^{2x}}$$

The failure case of the Particular Integral of a differential equation. Solve the second order differential equation
$$\frac{d^2y}{dx^2} - 4\frac{dy}{dx} + 3y = e^{3x} \quad \ldots (1)$$

To find the complementary function
$$\frac{d^2y}{dx^2} - 4\frac{dy}{dx} + 3y = 0.$$

The corresponding auxiliary equation is
$m^2 - 4m + 3 = 0$
$$m = \frac{4 \pm \sqrt{16 - 12}}{2} = \frac{4 \pm 2}{2}, m_1 = 3 \text{ and } m_2 = 1.$$

The complementary function solution is
$y = Ae^{3x} + Be^x$.

To find the particular integral. Let $y = ke^{3x}$, it is observed that this is already included in the complementary function.

$$\frac{dy}{dx} = 3ke^{3x} \quad \text{and} \quad \frac{d^2y}{dx^2} = 9ke^{3x}.$$

Substituting in equation (1)
$$9ke^{3x} - 4\left(3ke^{3x}\right) + 3\left(ke^{3x}\right) = e^{3x}.$$

Equating the coefficient of e^{3x}, $9k - 12k + 3k = 1$ from which we find that 0 is equal to 1 and therefore the solution is inconsistent. Therefore, when the right hand side is $e^{\alpha x}$ where α is a root of the auxiliary equation then we must use an alternative solution.

Let $y = kxe^{3x}$, $\dfrac{dy}{dx} = ke^{3x} + 3kxe^{3x}$ and $\dfrac{d^2y}{dx^2} = 3ke^{3x} + 3ke^{3x} + 9kxe^{3x}$ and substituting in equation (1), we have

$3ke^{3x} + 3ke^{3x} + 9kxe^{3x} - 4ke^{3x} - 12kxe^{3x} + 3kxe^{3x} = e^{3x}.$

Equating coefficients

$3k + 3k + 9kx - 4k - 12xk + 3kx = 1 \quad k = \dfrac{1}{2}.$

The general solution is $y = Ae^{3x} + Be^{x} + \dfrac{1}{2}xe^{3x}.$

It is observed that kx terms cancel as the k terms cancelled previously. If the k terms cancel in this case then we try a different solution.
The trial solution will be
$y = kx^2 e^{3x}.$

WORKED EXAMPLE 100

Solve the second order differential equations

(i) $\dfrac{d^2y}{dx^2} - 3\dfrac{dy}{dx} + 2y = 2e^{2x}$

(ii) $\dfrac{d^2y}{dx^2} - \dfrac{dy}{dx} - 6y = 3e^{-2x}$

(iii) $\dfrac{d^2y}{dx^2} + 8\dfrac{dy}{dx} + 16y = e^{-4x}$

(iv) $\dfrac{d^2y}{dx^2} - 2\dfrac{dy}{dx} + 5y = e^{x}.$

Solution 100

(i) $\dfrac{d^2y}{dx^2} - 3\dfrac{dy}{dx} + 2y = 2e^{2x}$ ...(1)

$m^2 - 3m + 2 = 0$ auxiliary equation

$m = \dfrac{3 \pm \sqrt{9-8}}{2} = \dfrac{3 \pm 1}{2}$

$m_1 = 2, \quad \text{or} \quad m_2 = 1$

$y = Ae^{2x} + Be^{x}$ the C.F.

The particular integral let $y = kx\, e^{2x}$

$\dfrac{dy}{dx} = ke^{2x} + 2kxe^{2x}$

$\dfrac{d^2y}{dx^2} = 2ke^{2x} + 2ke^{2x} + 2kx2e^{2x}$

substituting in (1)

$4ke^{2x} + 4kxe^{2x} - 3ke^{2x} - 6kxe^{2x} + 2kxe^{2x} = 2e^{2x}$

$ke^{2x} = 2e^{2x}$

$k = 2$

Therefore the P.I. solution is $y = 2xe^{2x}$ and the general solution = C.F. + P.I.

$\boxed{y = Ae^{2x} + Be^{x} + 2xe^{2x}}$

(ii) $\dfrac{d^2y}{dx^2} - \dfrac{dy}{dx} - 6y = 3e^{-2x}$ ...(2)

The C.F. can be found if $\dfrac{d^2y}{dx^2} - \dfrac{dy}{dx} - 6y = 0$, its auxiliary equation is $m^2 - m - 6 = 0,$

$m = \dfrac{1 \pm \sqrt{1+24}}{2} = \dfrac{1 \pm 5}{2},$

$m_1 = 3, \quad \text{or} \quad m_2 = -2.$

The C.F. is $y = Ae^{3x} + Be^{-2x}.$

The P.I. can be found if $y = kxe^{-2x},$

$\dfrac{dy}{dx} = ke^{-2x} - 2kxe^{-2x},$

$\dfrac{d^2y}{dx^2} = -2ke^{-2x} - 2ke^{-2x} + 4kxe^{-2x}.$

Substituting in (2)

$-2ke^{-2x} - 2ke^{-2x} + 4kxe^{-2x} - ke^{-2x} + 2kxe^{-2x} - 6kxe^{-2x} = 3e^{-2x}$

$-5ke^{-2x} = 3e^{-2x}, \; k = -\dfrac{3}{5},$ therefore the P.I.

$y = -\dfrac{3}{5}xe^{-2x}$ and the general solution is

$\boxed{y = Ae^{3x} + Be^{-2x} - \dfrac{3}{5}xe^{-2x}}$

(iii) $\dfrac{d^2y}{dx^2} + 8\dfrac{dy}{dx} + 16y = e^{-4x}$ ...(3)

The C.F. can be found if $\dfrac{d^2y}{dx^2} + 8\dfrac{dy}{dx} + 16y = 0,$ its auxiliary equation is $m^2 + 8m + 16 = 0.$

$(m+4)^2 = 0, m = -4,$ equal roots.

The C.F. is $y = e^{-4x}(A + Bx).$

The P.I. can be found if $y = kxe^{-4x},$

$\dfrac{dy}{dx} = -4kxe^{-4x} + ke^{-4x},$

$\dfrac{d^2y}{dx^2} = -4ke^{-4x} + 16kxe^{-4x} - 4ke^{-4x}.$

Substituting in (3) we have

$-4ke^{-4x} + 16xke^{-4x} - 4ke^{-4x} - 32kxe^{-4x} + 8ke^{-4x} + 16kxe^{-4x} = e^{-4x}, 0 = 1$, it is inconsistent, therefore we try $y = kx^2e^{-4x}$,

$$\frac{dy}{dx} = 2kxe^{-4x} - 4kx^2e^{-4x},$$

$$\frac{d^2y}{dx^2} = 2ke^{-4x} - 8kxe^{-4x} - 8kxe^{-4x} + 16kx^2e^{-4x}.$$

Substituting in (3) we have

$2ke^{-4x} - 8kxe^{-4x} - 8kxe^{-4x} + 16kx^2e^{-4x} + 16kxe^{-4x} - 32kx^2e^{-4x} + 16kx^2e^{-4x} = e^{-4x}$

$2k = 1, k = \dfrac{1}{2}$

therefore $y = \dfrac{1}{2}x^2e^{-4x}$ is the P.I.

The general solution is

$$\boxed{y = e^{-4x}(A + Bx) + \frac{1}{2}x^2e^{-4x}}$$

(iv) $\dfrac{d^2y}{dx^2} - 2\dfrac{dy}{dx} + 5y = e^x.$ ... (4)

The C.F. can be found if $\dfrac{d^2y}{dx^2} - 2\dfrac{dy}{dx} + 5y = 0$ its auxiliary equation is $m^2 - 2m + 5 = 0$. $m = 1 \pm i2$ and the C.F. is $y = e^x(A\cos 2x + B\sin 2x)$.

The particular integral.

If $y = ke^x$, $\dfrac{dy}{dx} = ke^x$, $\dfrac{d^2y}{dx^2} = ke^x$

$ke^x - 2ke^x + 5ke^x = e^x$

$4k = 1$

$k = \dfrac{1}{4}.$

The general solution

$$\boxed{y = e^x(A\cos 2x + B\sin 2x) + \frac{1}{4}e^x.}$$

Exercises 14

1. $\dfrac{d^2y}{dx^2} - y = 0$

2. $\dfrac{d^2y}{dx^2} - \dfrac{dy}{dx} - 12y = 0$

3. $\dfrac{d^2y}{dx^2} - 4\dfrac{dy}{dx} + 3y = 0$

4. $\dfrac{d^2y}{dx^2} + 4\dfrac{dy}{dx} + 4y = 0$

5. $\dfrac{d^2y}{dx^2} - 10\dfrac{dy}{dx} + 25y = 0$

6. $\dfrac{d^2y}{dx^2} - 2\dfrac{dy}{dx} + 2y = 0$

7. $\dfrac{d^2y}{dx^2} + 16\dfrac{dy}{dx} = 0$

8. $\dfrac{d^2y}{dx^2} + 4\dfrac{dy}{dx} + 29y = 0$

9. $\dfrac{d^2y}{dx^2} - \dfrac{dy}{dx} - 2y = 0$

10. $6\dfrac{d^2y}{dx^2} - \dfrac{dy}{dx} - 2y = 0$

11. $\dfrac{d^2y}{dx^2} - 4\dfrac{dy}{dx} + 4y = 0$

12. $\dfrac{d^2y}{dx^2} - 6\dfrac{dy}{dx} + 9y = 0$

13. $\dfrac{d^2y}{dx^2} + 8\dfrac{dy}{dx} + 16y = 0$

14. $\dfrac{d^2y}{dx^2} - 4\dfrac{dy}{dx} + 13y = 0$

15. $\dfrac{d^2y}{dx^2} + 6\dfrac{dy}{dx} + 10y = 0$

16. $9\dfrac{d^2y}{dx^2} + 6\dfrac{dy}{dx} + y = 0$

17-32. Find the particular solutions for the above second order linear differential equations with constant coefficients when $x = 0$, $y = 1$ and $\dfrac{dy}{dx} = -2$.

33. Solve the differential equation

$$5\frac{d^2x}{dt^2} + 14\frac{dx}{dt} - 3x = 4\sin t$$

given that at $t = 0$, $x = 0$ and $\frac{dx}{dt} = 2$.

34. Solve the differential equation

$$\frac{d^2x}{dt^2} + 12\frac{dx}{dy} + 36x = 0$$

given that at $t = 0$, $x = 0$ and $\frac{dx}{dt} = 1$.

35. Solve the differential equation

$$\frac{d^2z}{dx^2} + 16z = 7x^2 - x + 5.$$

36. Solve the differential equation

$$\frac{d^2y}{dx^2} - 9y = 15e^{5x}.$$

37. Solve the differential equation

$$\frac{d^2y}{dx^2} - \frac{dy}{dx} - 2y = 15.$$

38. Solve the differential equation

$$\frac{d^2y}{dx^2} + 6\frac{dy}{dx} + 8y = e^{-3x}$$

given that at $x = 0$, $y = 1$ and $\frac{dy}{dx} = 1$.

15

Length of Arc and Surface of Revolution

Length of Arc in Cartesian Coordinates

Consider a curve f(x) and two points.

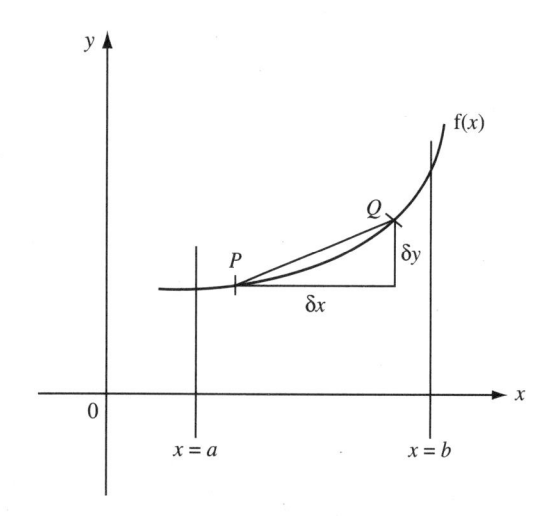

Fig. 7-I/41

P and Q are very close together, but for clarity are as shown in the figure.

From a right angled triangle, the vertical side is δy and the horizontal side δx, PQ is a chord and PQ also there is an arc, since the points P and Q are very close together we can approximate the chord PQ with that of the arc PQ.

$PQ^2 = (\delta x)^2 + (\delta y)^2$ using Pythagoras theorem

$\delta s = PQ$

$(\delta s)^2 = (\delta x)^2 + (\delta y)^2$

dividing each term by δx^2, we have

$$\frac{(\delta s)^2}{(\delta x)^2} = 1 + \frac{(\delta y)^2}{(\delta x)^2}$$

$$\left(\frac{\delta s}{\delta x}\right)^2 = 1 + \left(\frac{\delta y}{\delta x}\right)^2$$

as $\delta x \to 0$, $\frac{\delta y}{\delta x} \to \frac{dy}{dx}$ and hence $\frac{\delta s}{\delta x} \to \frac{ds}{dx}$
therefore

$$\left(\frac{ds}{dx}\right)^2 = 1 + \left(\frac{dy}{dx}\right)^2, ds = \sqrt{\left(1 + \left(\frac{dy}{dx}\right)^2\right)} dx.$$

Summing up all the elemental lengths of arcs we have

$$\int_a^b ds = \int_a^b \sqrt{\left(1 + \left(\frac{dy}{dx}\right)^2\right)} dx$$

$$s = \text{Length of arc} = \int_a^b \sqrt{\left(1 + \left(\frac{dy}{dx}\right)^2\right)} dx.$$

Length of Arc in Polar Coordinates

$(ds)^2 = (dx)^2 + (dy)^2$

$\cos\theta = \dfrac{x}{r}$ ...(1)

$\sin\theta = \dfrac{y}{r}$ ...(2)

$r^2 = x^2 + y^2.$

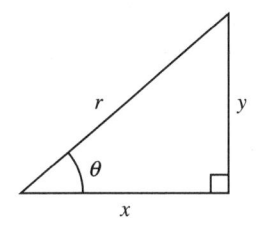

Fig. 7-I/42

Differentiating with respect to r

$x = r\cos\theta, \quad \dfrac{dx}{dr} = \cos\theta - r\sin\theta\,\dfrac{d\theta}{dr}$

$y = r\sin\theta, \quad \dfrac{dy}{dr} = \sin\theta + r\cos\theta\,\dfrac{d\theta}{dr}$

$(dx)^2 + (dy)^2 = (dr\cos\theta - r\sin\theta\,d\theta)^2$

$\qquad\qquad\qquad + (dr\sin\theta + r\cos\theta\,d\theta)^2$

$\qquad = (dr)^2\cos^2\theta - 2r\sin\theta\cos\theta\,dr\,d\theta$

$\qquad\quad + r^2\sin^2\theta(d\theta)^2 + (dr)^2\sin^2\theta$

$\qquad\quad + 2r\sin\theta\cos\theta\,dr\,d\theta$

$\qquad\quad + r^2\cos^2\theta(d\theta)^2$

$\qquad = (dr)^2(\sin^2\theta + \cos^2\theta)$

$\qquad\quad + r^2(\cos^2\theta + \sin^2\theta)(d\theta)^2$

$(ds)^2 = (dr)^2 + r^2(d\theta)^2$

dividing each term by $(d\theta)^2$

$\left(\dfrac{ds}{d\theta}\right)^2 = \left(\dfrac{dr}{d\theta}\right)^2 + r^2.$

Length or arc $= s$

$\qquad = \displaystyle\int_{\theta_1}^{\theta_2}\sqrt{\left(\dfrac{dr}{d\theta}\right)^2 + r^2}\,d\theta.$

Length of Arc Using Parametric Equations

Length of arc $= s = \displaystyle\int_a^b \sqrt{\left(1 + \left(\dfrac{dy}{dx}\right)^2\right)}\,dx$

$\left(\dfrac{dy}{dx}\right)^2 = \left(\dfrac{\frac{dy}{dt}}{\frac{dx}{dt}}\right)^2 = \left(\dfrac{\dot y}{\dot x}\right)^2$

$s = \displaystyle\int_{t_1}^{t_2}\sqrt{1 + \left(\dfrac{\dot y}{\dot x}\right)^2}\,dx$

$\quad = \displaystyle\int_{t_1}^{t_2}\sqrt{(\dot x^2 + \dot y^2)}\,dt$

$\quad = \displaystyle\int_{t_1}^{t_2}\sqrt{\left[\left(\dfrac{dx}{dt}\right)^2 + \left(\dfrac{dy}{dt}\right)^2\right]}\,dt.$

WORKED EXAMPLE 101

A curve is given by the parametric equation $x = \sin t$ and $y = \cos t$. Find the length of arc between $x = 0$ and $x = 1$.

Solution 101

$x = \sin t, \quad \dfrac{dx}{dt} = \cos t$

$y = \cos t, \quad \dfrac{dy}{dt} = -\sin t.$

Length of arc $= \displaystyle\int_0^{\frac{\pi}{2}}\sqrt{(\cos t)^2 + (-\sin t)^2}\,dt$

when $x = 0, t = 0$;

when $x = 1, t = \dfrac{\pi}{2}$;

$= \displaystyle\int_0^{\frac{\pi}{2}}\sqrt{\cos^2 t + \sin^2 t}\,dt$

$= \displaystyle\int_0^{\frac{\pi}{2}}dt = \big[t\big]_0^{\frac{\pi}{2}} = \dfrac{\pi}{2}.$

WORKED EXAMPLE 102

Show that the circumference of a circle of radius r is given by $C = 2\pi r$.

Solution 102

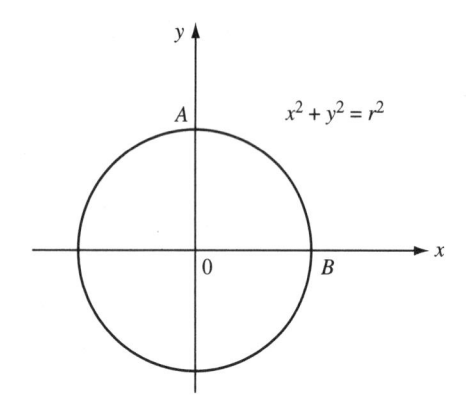

$$x^2 + y^2 = r^2 \qquad \ldots(1)$$

Fig. 7-I/43

Differentiating equation (1) with respect to x, $2x + 2y\dfrac{dy}{dx} = 0$, $\dfrac{dy}{dx} = -\dfrac{2x}{2y} = -\dfrac{x}{y}$

Length of arc $AB = \displaystyle\int_0^r \sqrt{\left[1 + \left(\dfrac{dy}{dx}\right)^2\right]}\, dx$

$= \displaystyle\int_0^r \sqrt{\left[1 + \left(-\dfrac{x}{y}\right)^2\right]}\, dx$

$= \displaystyle\int_0^r \sqrt{\left(1 + \dfrac{x^2}{y^2}\right)}\, dx$

$= \displaystyle\int_0^r \sqrt{\dfrac{y^2 + x^2}{y^2}}\, dx = \int_0^r \dfrac{r}{y}\, dx$

$= \displaystyle\int_0^r \dfrac{r}{\sqrt{r^2 - x^2}}\, dx$

$= \displaystyle\int_0^r \dfrac{r}{\sqrt{r^2 - r^2 \sin^2 \theta}}\, r\cos\theta\, d\theta.$

Let $x = r \sin\theta$, $\dfrac{dx}{d\theta} = r \cos\theta$

if $x = r$, $\sin\theta = 1$, $\theta = \dfrac{\pi}{2}$

if $x = 0$, $\sin\theta = 0$, $\theta = 0$.

Length of arc $AB = \displaystyle\int_0^{\frac{\pi}{2}} r\, d\theta = \Big[r\theta\Big]_0^{\frac{\pi}{2}} = r\dfrac{\pi}{2}.$

The circumference is $4 \times \dfrac{r\pi}{2} = 2\pi r.$

$$\boxed{C = 2\pi r}$$

WORKED EXAMPLE 103

Find the length of arc of the curve $y = \cosh x$ between $x = 0$ and $x = 2$.

Solution 103

$y = \cosh x$, $\dfrac{dy}{dx} = \sinh x$

Length of arc $= \displaystyle\int_0^2 \sqrt{\left[1 + \left(\dfrac{dy}{dx}\right)^2\right]}\, dx$

$= \displaystyle\int_0^2 \sqrt{\left(1 + \sinh^2 x\right)}\, dx$

$= \displaystyle\int_0^2 \cosh x\, dx$

since $\cosh^2 x = \sinh^2 x + 1$

$= \displaystyle\int_0^2 \cosh x\, dx = \Big[\sinh x\Big]_0^2 = \sinh 2$

$= 3.63$ units.

WORKED EXAMPLE 104

Find the length of arc of the curve $x = t + \sin t$, $y = 1 + \cos t$ between $t = 0$ and $t = \dfrac{\pi}{2}$.

Solution 104

$x = t + \sin t, \quad \dfrac{dx}{dt} = 1 + \cos t = \dot{x}$

$y = 1 + \cos t, \quad \dfrac{dy}{dt} = -\sin t = \dot{y}$

Length of arc $= \displaystyle\int_0^{\frac{\pi}{2}} \sqrt{(\dot{x}^2 + \dot{y}^2)}\, dt$

$\dot{x} = \dfrac{dx}{dt}, \quad \dot{y} = \dfrac{dy}{dt}$

$= \displaystyle\int_0^{\frac{\pi}{2}} \sqrt{\left(1 + 2\cos t + \cos^2 t + \sin^2 t\right)}\, dt$

$= \displaystyle\int_0^{\frac{\pi}{2}} \sqrt{2 + 2\cos t}\, dt$

$= \sqrt{2}\displaystyle\int_0^{\frac{\pi}{2}} \sqrt{1 + \cos t}\, dt$

$\cos t = 2\cos^2 \dfrac{t}{2} - 1$ hence $\cos t + 1 = 2\cos^2 \dfrac{t}{2}$

$= \sqrt{2}\displaystyle\int_0^{\frac{\pi}{2}} \sqrt{2\cos^2 \dfrac{t}{2}}\, dt$

$= 2\displaystyle\int_0^{\frac{\pi}{2}} \cos \dfrac{t}{2}\, dt$

$= 2\left[\dfrac{\sin \frac{t}{2}}{\frac{1}{2}}\right]_0^{\frac{\pi}{2}}$

$= 4 \sin \dfrac{\pi}{4}$

$= 4 \dfrac{1}{\sqrt{2}} \dfrac{\sqrt{2}}{\sqrt{2}}$

$= 2\sqrt{2}.$

WORKED EXAMPLE 105

Calculate the length of arc of the curve $y = 3e^{3x}$ from $x = 1$ to $x = 2$, use Simpson's Rule for five ordinates.

Solution 105

$y = 3e^{3x} \quad \dfrac{dy}{dx} = 9e^{3x}$

$s = \displaystyle\int_1^2 \sqrt{\left[1 + \left(\dfrac{dy}{dx}\right)^2\right]}\, dx = \displaystyle\int_1^2 \sqrt{\left[1 + \left(9e^{3x}\right)^2\right]}\, dx$

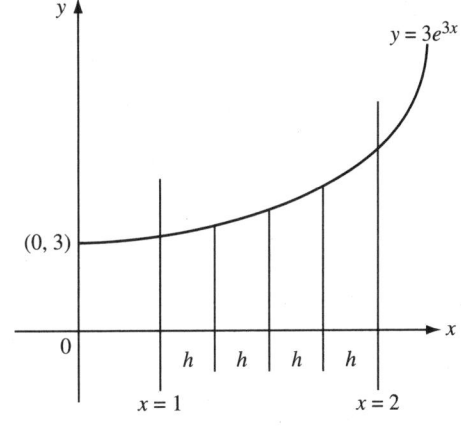

$= \displaystyle\int_1^2 \sqrt{\left(1 + 81e^{6x}\right)}\, dx$

Fig. 7-I/44

$h = \dfrac{2 - 1}{4} = \dfrac{1}{4} = 0.25$

$\displaystyle\int_1^2 \sqrt{\left(1 + 81e^{6x}\right)}\, dx$

$\approx \dfrac{h}{3}[y_1 + y_5 + 4(y_2 + y_4) + 2y_3]$

$= \dfrac{0.25}{3}[181 + 3631 + 4(383 + 1715) + 2 \times 810]$

$= \dfrac{0.25}{3}[181 + 3631 + 8392 + 1620] = 1152.$

x	1	1.25	1.50	1.75	2.00
$y = \sqrt{1 + 81e^{6x}}$	181	383	810	1715	3631

WORKED EXAMPLE 106

Determine the length of arc of a curve given by the parametric equations $x = t - \sin t$, $y = 1 - \cos t$ between $t = 0$ and $t = \pi$.

Solution 106

Consider now a curve which is given parametrically

$x = t - \sin t$
$y = 1 - \cos t$.

Differentiating with respect to t,

$$\frac{dx}{dt} = 1 - \cos t, \quad \frac{dy}{dt} = \sin t$$

$$\sqrt{\left[1 + \left(\frac{dy}{dx}\right)^2\right]} dx = \sqrt{\left[1 + \left(\frac{\frac{dy}{dt}}{\frac{dx}{dt}}\right)^2\right]} dx$$

$$= \sqrt{\left(\frac{dx}{dt}\right)^2 + \left(\frac{dy}{dt}\right)^2} \frac{dx}{\frac{dx}{dt}}$$

$$= \sqrt{(\dot{x}^2 + \dot{y}^2)} \, dt$$

The length of arc $= \int_{t_1}^{t_2} \sqrt{(\dot{x}^2 + \dot{y}^2)} \, dt$

when $\dot{x} = \frac{dx}{dt}, \quad \dot{y} = \frac{dy}{dt}$.

Length of arc $= \int_0^{\pi} \sqrt{(\dot{x}^2 + \dot{y}^2)} \, dt$

$$= \int_0^{\pi} \sqrt{(1 - \cos t)^2 + \sin^2 t} \, dt$$

$$= \int_0^{\pi} \sqrt{\left(1 - 2\cos t + \cos^2 t + \sin^2 t\right)} \, dt$$

$$= \int_0^{\pi} \sqrt{2(1 - \cos t)} \, dt$$

$$= \int_0^{\pi} \sqrt{2\left(1 - \left(2\cos^2 \frac{t}{2} - 1\right)\right)} \, dt$$

$$= \int_0^{\pi} 2\sqrt{1 - \cos^2 \frac{t}{2}} \, dt = \int_0^{\pi} 2 \sin \frac{t}{2} \, dt$$

$$= 2\left[-\cos \frac{t}{2}\right]_0^{\pi} = 2\left[-\cos \frac{\pi}{2} + \cos 0\right]$$

$= 2$ units.

WORKED EXAMPLE 107

Find the length of arc of the curve with coordinates $(\sin 2t, \cos 2t)$ between $t = 0$ and $t = \pi$.

Solution 107

$x = \sin 2t, \quad \dfrac{dx}{dt} = 2 \cos 2t$

$y = \cos 2t, \quad \dfrac{dy}{dt} = -2 \sin 2t$.

Length of arc $= \int_0^{\pi} \sqrt{(\dot{x}^2 + \dot{y}^2)} \, dt$

$$= \int_0^{\pi} \sqrt{(2 \cos 2t)^2 + (-2 \sin 2t)^2} \, dt$$

$$= \int_0^{\pi} \sqrt{4 \cos^2 2t + 4 \sin^2 2t} \, dt$$

$$= \int_0^{\pi} 2 \, dt = \left[2t\right]_0^{\pi} = 2\pi \text{ units.}$$

WORKED EXAMPLE 108

Calculate the length of the arc of the curve whose parametric equations are

$x = 3 \cos t - \cos 3t$
$y = 3 \sin t - \sin 3t$

between the points corresponding to $t = 0$ and $t = \dfrac{\pi}{4}$.

Solution 108

Length of arc $= \int_0^{\pi/4} \sqrt{\dot{x}^2 + \dot{y}^2} \, dt$

$x = 3 \cos t - \cos 3t$,

$\dfrac{dx}{dt} = -3 \sin t + 3 \sin 3t = \dot{x}$

$y = 3 \sin t - \sin 3t$,

$\dfrac{dy}{dt} = +3 \cos t - 3 \cos 3t = \dot{y}$

Length of Arc and Surface of Revolution

Length of arc

$$= \int_0^{\frac{\pi}{4}} \sqrt{\begin{pmatrix} 9\sin^2 t + 9\sin^2 3t - 18\sin t \sin 3t + \\ 9\cos^2 t + 9\cos^2 3t - 18\cos t \cos 3t \end{pmatrix}} \, dt$$

$$= \int_0^{\frac{\pi}{4}} \sqrt{9 + 9 - 18(\cos t \cos 3t + \sin t \sin 3t)} \, dt$$

$$= \sqrt{18} \int_0^{\frac{\pi}{4}} \sqrt{1 - \cos 2t} \, dt$$

$$= \sqrt{18} \int_0^{\frac{\pi}{4}} \sqrt{1 - 1 + 2\sin^2 t} \, dt$$

$\cos 2t = 1 - 2\sin^2 t$

$$= \sqrt{18} \sqrt{2} \int_0^{\frac{\pi}{4}} \sin t \, dt$$

$$= 6 \int_0^{\frac{\pi}{4}} \sin t \, dt = \left[-6\cos t\right]_0^{\frac{\pi}{4}}$$

$$= -\frac{6\sqrt{2}}{2} + 6 = (6 - 3\sqrt{2}) \text{ units.}$$

WORKED EXAMPLE 109

Sketch the graph $x^2 = y$. Find the length of this curve between the limits $x = -3$ and $x = -1$.

Solution 109

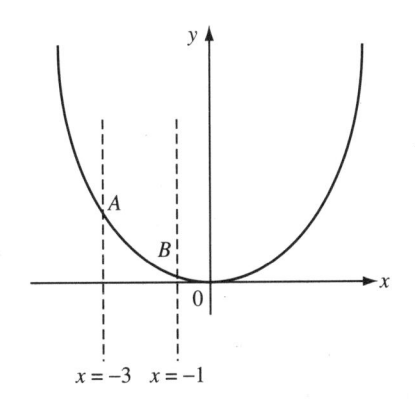

Fig. 7-I/45

Length of arc $AB = \int_{-3}^{-1} \sqrt{\left[1 + \left(\frac{dy}{dx}\right)^2\right]} \, dx$

$$= \int_{-3}^{-1} \sqrt{1 + 4x^2} \, dx$$

$x^2 = y$, differentiating with respect to x

$2x = \frac{dy}{dx}, \frac{dy}{dx} = 2x.$

Length of arc $AB = \int_{-3}^{-1} \sqrt{(1 + 4x^2)} \, dx$

Let $2x = \sinh y$, $2\frac{dx}{dy} = \cosh y$

when $x = -1$, $\sinh y = -2$, $y = -1.44$
when $x = -3$, $\sinh y = -6$, $y = -2.49$.

Length of arc $AB = \int_{-2.49}^{-1.44} \sqrt{1 + \sinh^2 y} \, \frac{\cosh y}{2} \, dy$

$$= \frac{1}{2} \int_{-2.49}^{-1.44} \cosh^2 y \, dy$$

$\cosh 2y = 2\cosh^2 y - 1,$

$\frac{\cosh 2y + 1}{2} = \cosh^2 y$

$\frac{1}{4} \int_{-2.49}^{-1.44} (\cosh 2y + 1) dy = \frac{1}{4} \left(\frac{\sinh 2y}{2} + y\right)_{-2.49}^{-1.44}$

$$= \frac{1}{4}(-4.44 - 1.44 + 36.4 + 2.49)$$

Length of arc $AB = 8.25$ units.

WORKED EXAMPLE 110

Find the locus of the curve whose parametric equations are $x = 4t^2$, $y = 4t$. Hence find the length of the curve from $A(1, 2)$ to $B(4, 4)$.

Solution 110

$x = 4t^2, y = 4t, t = \dfrac{y}{4}, x = 4\left(\dfrac{y}{4}\right)^2$

$\boxed{y^2 = 4x}$ the locus.

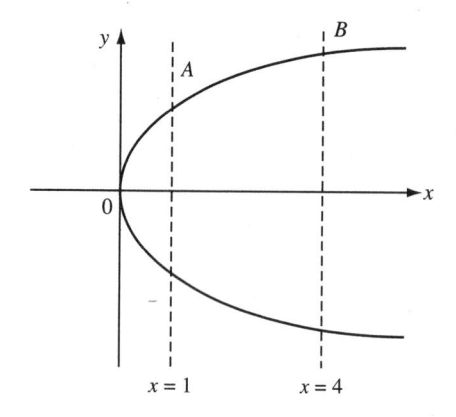

Fig. 7-I/46

Length of arc $= \displaystyle\int_1^4 \sqrt{1 + \left(\dfrac{dy}{dx}\right)^2}\, dx$

$2y\dfrac{dy}{dx} = 4, \quad \dfrac{dy}{dx} = \dfrac{2}{y}, \quad \left(\dfrac{dy}{dx}\right)^2 = \dfrac{4}{y^2}$

$\displaystyle\int_1^4 \sqrt{1 + \dfrac{4}{y^2}}\, dx = \int_1^4 \sqrt{1 + \dfrac{4}{4x}}\, dx$

$= \displaystyle\int_1^4 \sqrt{1 + \dfrac{1}{x}}\, dx$

$\displaystyle\int_1^4 \sqrt{1 + \dfrac{1}{x}}\, dx = \int_{\frac{1}{2}}^1 \sqrt{1 + \dfrac{1}{4t^2}}\, 8t\, dt$

$= \displaystyle\int_{\frac{1}{2}}^1 \dfrac{\sqrt{4t^2 + 1}}{2t}\, 8t\, dt$

$x = 4t^2, \dfrac{dx}{dt} = 8t, x = 1, t = \dfrac{1}{2}; x = 4, t = 1$

$\displaystyle\int_{\frac{1}{2}}^1 4\sqrt{4t^2 + 1}\, dt \quad$ let $2t = \sinh y, 2\dfrac{dt}{dy} = \cosh y$

$\displaystyle\int_{\sinh^{-1} 1}^{\sinh^{-1} 2} 4\sqrt{\sinh^2 y + 1}\, \cosh y\, \dfrac{dy}{2}$

$= \displaystyle\int 2\cosh^2 y\, dy$

$\displaystyle\int_{\sinh^{-1} 1}^{\sinh^{-1} 2} (\cosh 2y + 1)\, dy = \left(\dfrac{\sinh 2y}{2} + y\right)_{\sinh^{-1} 1}^{\sinh^{-1} 2}$

$= \dfrac{\sinh 2 \sinh^{-1} 2}{2} + \sinh^{-1} 2$

$\quad - \dfrac{\sinh 2 \sinh^{-1} 1}{2}$

$\quad - \sinh^{-1} 1$

$= 4.47 + 1.44 - 1.414$

$\quad - 0.881 = 3.62$

$AB = 3.62$ units.

Surfaces

Area of surface of solid of revolution about the x-axis.

$= \displaystyle\int_{x_1}^{x_2} 2\pi y \sqrt{\left[1 + \left(\dfrac{dy}{dx}\right)^2\right]}\, dx$

$= \displaystyle\int_{t_1}^{t_2} 2\pi y \sqrt{\left[\left(\dfrac{dx}{dt}\right)^2 + \left(\dfrac{dy}{dt}\right)^2\right]}\, dt$

Area of surface of solid of revolution about the y-axis.

$= \displaystyle\int_{x_1}^{x_2} 2\pi x \sqrt{\left[1 + \left(\dfrac{dy}{dx}\right)^2\right]}\, dx$

$= \displaystyle\int_{t_1}^{t_2} 2\pi x \sqrt{\left[\left(\dfrac{dx}{dt}\right)^2 + \left(\dfrac{dy}{dt}\right)^2\right]}\, dx.$

An astroid is given by the Fig. 7-I/47 with the parametric equations.

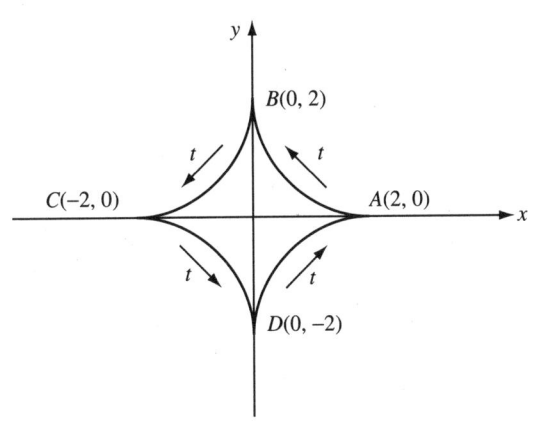

Fig. 7-I/47

$x = 2\cos^3 t$
$y = 2\sin^3 t$
$0 \leq t \leq 2\pi$.

(a) Determine the length of arc $0 \leq t \leq \dfrac{\pi}{2}$, hence find the perimeter of the astroid.

(b) The arc AB is rotated about the x-axis, find the surface area so formed and hence find the surface area of the astroid.

(a) $\displaystyle\int_0^{\frac{\pi}{2}} \sqrt{\left(\dfrac{dx}{dt}\right)^2 + \left(\dfrac{dy}{dt}\right)^2}\, dt$

$x = 2\cos^3 t,\ \dfrac{dx}{dt} = 6\cos^2 t(-\sin t)$

$y = 2\sin^3 t,\ \dfrac{dy}{dt} = 6\sin^2 t \cos t$

$= \displaystyle\int_0^{\frac{\pi}{2}} \sqrt{36\sin^2 t\, \cos^4 t + 36\sin^4 t\, \cos^2 t}\, dt$

$= 6\displaystyle\int_0^{\frac{\pi}{2}} \sin t\, \cos t\sqrt{\cos^2 t + \sin^2 t}\, dt$

$= \displaystyle\int_0^{\frac{\pi}{2}} 3\sin 2t\, dt = \left(-\dfrac{3}{2}\cos 2t\right)_0^{\frac{\pi}{2}}$

$= -\dfrac{3}{2}\cos\pi + \dfrac{3}{2}\cos 0 = 3$ units.

The perimeter of the astroid $= 4 \times 3 = 12$ units.

(b) $\displaystyle\int_0^{\frac{\pi}{2}} 2\pi y(3\sin 2t)\, dt = \displaystyle\int_0^{\frac{\pi}{2}} 2\pi\, 2\sin^3 t\, 3\sin 2t\, dt$

$= 2 \times 12\pi \displaystyle\int_0^{\frac{\pi}{2}} \sin^4 t \cos t\, dt$

$= 24\pi \displaystyle\int_0^{\frac{\pi}{2}} \sin^4 t\, d(\sin t)$

$= 24\pi \left(\dfrac{\sin^5 t}{5}\right)_0^{\frac{\pi}{2}} = \dfrac{24\pi}{5}$.

Total surface area $= \dfrac{48\pi}{5}$.

WORKED EXAMPLE 111

A curve is given by the parametric equations

$x = t + \sin t$

$y = 1 - \cos t$.

Sketch the curve for values of t from 0 to π.

Determine the length of arc between $t = 0$ and $t = \pi$. The length of this arc is rotated about the x-axis determine the surface area so generated.

Solution 111

$x = t + \sin t$ ...(1)

$y = 1 - \cos t$ ...(2)

when $t = 0,\ x = 0,\ y = 0$

$t = \dfrac{\pi}{4} \quad x = \dfrac{\pi}{4} + 0.707,\ y = 1 - 0.707$

$\quad x = 1.49,\quad y = 0.293$

$t = \dfrac{\pi}{2} \quad x = \dfrac{\pi}{2} + 1,\quad y = 1$

$\quad x = 2.57,\quad y = 1$

$t = \dfrac{3\pi}{4} \quad x = \dfrac{3\pi}{4} + 0.707,\ y = 1.707$

$\quad x = 3.063,\quad y = 1.707$

$t = \pi \quad x = \pi,\quad y = 2$

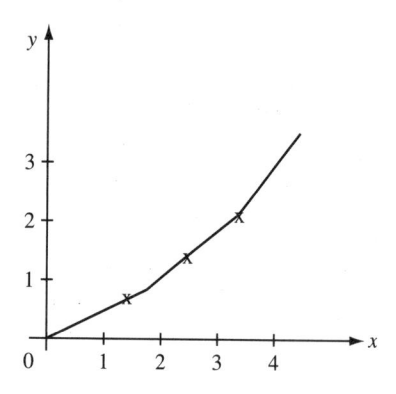

Fig. 7-I/48

Differentiate (1) and (2) with respect to x

$$\frac{dx}{dt} = 1 + \cos t$$

$$\frac{dy}{dt} = \sin t.$$

Length of arc $= \displaystyle\int_0^\pi \sqrt{\left[\left(\frac{dx}{dt}\right)^2 + \left(\frac{dy}{dt}\right)^2\right]}\, dt$

$= \displaystyle\int_0^\pi \sqrt{(1 + \cos t)^2 + \sin^2 t}\, dt$

$= \displaystyle\int_0^\pi \sqrt{(1 + 2\cos t + 1)}\, dt$

$= \displaystyle\int_0^\pi \sqrt{2(1 + \cos t)}\, dt$

$\cos t = 2\cos^2 \dfrac{t}{2} - 1$

$= \displaystyle\int_0^\pi \sqrt{2\left(2\cos^2 \dfrac{t}{2}\right)}\, dt$

$= \displaystyle\int_0^\pi 2\cos \dfrac{t}{2}\, dt$

$= \left[4\sin \dfrac{t}{2}\right]_0^\pi$

$= 4 \sin \dfrac{\pi}{2}$

$= 4$ units.

Surface area

$= \displaystyle\int_0^\pi 2\pi y \left(2\cos \dfrac{t}{2}\right) dt$

$= \displaystyle\int_0^\pi 2\pi(1 - \cos t) 2\cos \dfrac{t}{2}\, dt$

$= 4\pi \displaystyle\int_0^\pi \cos \dfrac{t}{2}\, dt - 4\pi \displaystyle\int_0^\pi \cos t \cos \dfrac{t}{2}\, dt$

$= \left[\dfrac{4\pi \sin \dfrac{t}{2}}{\dfrac{1}{2}}\right]_0^\pi - 4\pi \displaystyle\int_0^\pi \dfrac{1}{2} \cos \dfrac{t}{2}\, dt$

$\quad - 4\pi \displaystyle\int_0^\pi \dfrac{1}{2} \cos \dfrac{3t}{2}\, dt$

since $\cos t \cos \dfrac{t}{2} = \dfrac{1}{2}\left[\cos\left(t - \dfrac{t}{2}\right) + \cos\left(t + \dfrac{t}{2}\right)\right]$

$= \dfrac{1}{2} \cos \dfrac{t}{2} + \dfrac{1}{2} \cos \dfrac{3t}{2}$

$= 8\pi \sin \dfrac{\pi}{2} - 2\pi \left(\dfrac{\sin \dfrac{t}{2}}{\dfrac{1}{2}}\right)_0^\pi$

$\quad - 2\pi \left(\dfrac{\sin \dfrac{3t}{2}}{\dfrac{3}{2}}\right)_0^\pi$

$= 8\pi - 4\pi - \dfrac{4\pi}{3} \sin \dfrac{3\pi}{2}$

$= 4\pi + \dfrac{4\pi}{3}$

$= \dfrac{16\pi}{3}$ square units.

WORKED EXAMPLE 112

Find the length of arc of the curve given by the parametric equations.

$$x = 2t^2$$
$$y = 2t^3$$

for $1 \leq t \leq 2$.

Solution 112

Length of arc $= \int_1^2 \sqrt{\left(\dfrac{dx}{dt}\right)^2 + \left(\dfrac{dy}{dt}\right)^2}\, dt$

$= \int_1^2 \sqrt{(4t)^2 + (6t^2)^2}\, dt$

$x = 2t^2, \dfrac{dx}{dt} = 4t \qquad y = 2t^3, \dfrac{dy}{dt} = 6t^2$

$= \int_1^2 \sqrt{16t^2 + 36t^4}\, dt$

$= \int_1^2 4t\sqrt{1 + \dfrac{36}{16}t^2}\, dt$

$= \int_1^2 4t\sqrt{1 + \dfrac{9}{4}t^2}\, dt \text{ let } \dfrac{3}{2}t = \tan\theta, \dfrac{3}{2}\dfrac{dt}{d\theta} = \sec^2\theta$

$= \int_{\tan^{-1} 1.5}^{\tan^{-1} 3} 4\left(\dfrac{2\tan\theta}{3}\right)\sqrt{(1+\tan^2\theta)}\dfrac{2}{3}\sec^2\theta\, d\theta$

$= \dfrac{16}{9}\int_{\tan^{-1} 1.5}^{\tan^{-1} 3} \tan\theta \sec\theta \sec^2\theta\, d\theta$

$= \dfrac{16}{9}\int_{\tan^{-1} 1.5}^{\tan^{-1} 3} \dfrac{\sin\theta}{\cos^4\theta}\, d\theta$

$= \dfrac{16}{9}\int_{\tan^{-1} 1.5}^{\tan^{-1} 3} -\dfrac{d(\cos\theta)}{\cos^4\theta}$

$= \dfrac{16}{9}\left(-\dfrac{(\cos\theta)^{-3}}{-3}\right)_{\tan^{-1} 1.5}^{\tan^{-1} 3}$

$= \left(\dfrac{16}{27\cos^3\theta}\right)_{\tan^{-1} 1.5}^{\tan^{-1} 3}$

$\sqrt{4+9t^2}$, $3t$, θ, 2

Fig. 7-I/49

From Fig. 7-I/49

$\cos\theta = \dfrac{2}{\sqrt{4+9t^2}}$

$= \left[\dfrac{16}{27}\left(\dfrac{\sqrt{4+9t^2}}{2}\right)^3\right]_1^2$

$= \left[\dfrac{2}{27}(4+9t^2)^{\frac{3}{2}}\right]_1^2$

$= \dfrac{2}{27}\left(40^{\frac{3}{2}} - 13^{\frac{3}{2}}\right)$

$= \dfrac{2}{27}(252.98 - 46.87)$

$= 15.3$ units.

WORKED EXAMPLE 113

The length of arc of the curve given by the parametric equations

$x = 2t^2$

$y = 2t^3$

for $1 \leq t \leq 2$, is revolved about the y-axis, determine the surface of revolution for these limits.

Solution 113

Surface of revolution

$= \int_1^2 2\pi x\sqrt{\left(\dfrac{dx}{dt}\right)^2 + \left(\dfrac{dy}{dt}\right)^2}\, dt$

$= \int_1^2 2\pi\, 2t^2\sqrt{(4t)^2 + (6t^2)^2}\, dt$

$= 16\pi \int_1^2 t^3\sqrt{1 + \dfrac{9}{4}t^2}\, dt$

let $\dfrac{3}{2}t = \tan\theta,\ \dfrac{3}{2}\dfrac{dt}{d\theta} = \sec^2\theta$

$$= 16\pi \int_{\tan^{-1} 1.5}^{\tan^{-1} 3} \left(\frac{8}{27}\tan^3\theta\right)\sqrt{1+\tan^2\theta}\,\frac{2\,d\theta\,\sec^2\theta}{3}$$

$$= \frac{256}{81}\pi \int_{\tan^{-1} 1.5}^{\tan^{-1} 3} \tan^3\theta \sec^3\theta\,d\theta$$

$$= \frac{256}{81}\pi \int_{\tan^{-1} 1.5}^{\tan^{-1} 3} \frac{\sin^3\theta}{\cos^6\theta}\,d\theta$$

$$= \frac{256}{81}\pi \int_{\tan^{-1} 1.5}^{\tan^{-1} 3} \frac{\sin^2\theta}{\cos^6\theta}\,d(-\cos\theta)$$

$$= \frac{256}{81}\pi \int_{\tan^{-1} 1.5}^{\tan^{-1} 3} \frac{1-\cos^2\theta}{\cos^6\theta}\,d(-\cos\theta)$$

$$= \frac{256}{81}\pi \int_{\tan^{-1} 1.5}^{\tan^{-1} 3} \left[(\cos\theta)^{-6} - (\cos\theta)^{-4}\right] d(-\cos\theta)$$

$$= \frac{256}{81}\pi \left[\frac{(\cos\theta)^{-5}}{-5} - \frac{(\cos\theta)^{-3}}{-3}\right]_{\tan^{-1} 1.5}^{\tan^{-1} 3} (-1)$$

$$= \frac{256}{81}\pi \left[\frac{1}{5\cos^5\theta} - \frac{1}{3\cos^3\theta}\right]_{\tan^{-1} 1.5}^{\tan^{-1} 3}$$

$$= \frac{256}{81}\pi \left[\frac{(4+9t^2)^{\frac{5}{2}}}{5\times 32} - \frac{(4+9t^2)^{\frac{3}{2}}}{3\times 8}\right]_{1}^{2}$$

$$= \frac{256}{81}\pi[63.3 - 3.81 - 10.5 + 1.95]$$

$$= 506 \text{ square units.}$$

Exercises 15

1. Determine the length of arc of the curve $y = 2x^2$ between $x = 1$ and $x = 2$.

2. Determine the length of the perimeter of the astroid curve $x = 3\cos^3 t$, $y = 3\sin^3 t$.

3. Find the length of the circumference of $r = 1 - \sin\theta$.

4. Find the length of the arc of the parabola $y^2 = 4x$ between $x = 0$ and $x = 3$.

5. Find the length of the arc of the cycloid $x = \theta - \sin\theta$, $y = 1 - \cos\theta$ for $0 \leq \theta \leq \pi$.

6. Find the length of the arc of the curve with parametric equations $x = \tanh t$, $y = \operatorname{sech} t$ when $0 \leq t \leq 1$.

7. Find the length of the arc of the curves with polar coordinates.
   (i) $r = \sin\theta$,
   (ii) $r = \cos\theta$, $0 \leq \theta \leq \frac{\pi}{2}$.

8. A region bounded by the x-axis, the ordinates $x = 1$ and $x = 3$ and the arc of the parabola $y = 4\sqrt{x}$ between these ordinates, is rotated through 2π radians about the x-axis. Determine the surface area of the solid so formed.

9. Determine the length of the arc of the curve $y = 4x^2$ between $x = 0$ and $x = 2$.

10. The parametric equations of a curve are
    $$x = 2\cos t - \cos 2t$$
    $$y = 2\sin t - \sin 2t.$$
    Find the length of the arc of the curve for which $0 \leq t \leq \frac{\pi}{2}$.

    Find also the surface formed when this arc is revolved through 2π radians about the x-axis.

11. Find the length of the arc of the curve $y^2 = x^3$ between $x = 1$ and $x = 2$.

16

Integration Using t Formulae

Let $t = \tanh \dfrac{x}{2}$, $\dfrac{dt}{dx} = \dfrac{1}{2}\operatorname{sech}^2 \dfrac{x}{2}$,

$$dx = \dfrac{2\,dt}{\operatorname{sech}^2 \dfrac{x}{2}} = \dfrac{2\,dt}{1 - t^2}$$

$$1 - \tanh^2 \dfrac{x}{2} = \operatorname{sech}^2 \dfrac{x}{2}$$

$$1 - t^2 = \operatorname{sech}^2 \dfrac{x}{2}$$

$$\sinh x = 2\sinh \dfrac{x}{2} \cosh \dfrac{x}{2}$$

$$= \dfrac{2 \sinh \dfrac{x}{2} \cosh \dfrac{x}{2}}{\cosh^2 \dfrac{x}{2} - \sinh^2 \dfrac{x}{2}} = \dfrac{2 \tanh \dfrac{x}{2}}{1 - \tanh^2 \dfrac{x}{2}}$$

$$\sinh x = \dfrac{2t}{1 - t^2}$$

$$\cosh x = \dfrac{1 + t^2}{1 - t^2}, \quad \tanh x = \dfrac{2t}{1 + t^2}$$

$$\operatorname{cosech} x = \dfrac{1 - t^2}{2t}, \quad \coth x = \dfrac{1 + t^2}{2t}$$

$$\operatorname{sech} x = \dfrac{1 - t^2}{1 + t^2}, \quad \sinh x = \dfrac{2t}{1 - t^2}.$$

From circular functions

$$\sin x = \dfrac{2t}{1 + t^2}, \quad \operatorname{cosec} x = \dfrac{1 + t^2}{2t}$$

$$\cos x = \dfrac{1 - t^2}{1 + t^2}, \quad \sec x = \dfrac{1 + t^2}{1 - t^2}$$

$$\tan x = \dfrac{2t}{1 + t^2}, \quad \cot x = \dfrac{1 + t^2}{2t}$$

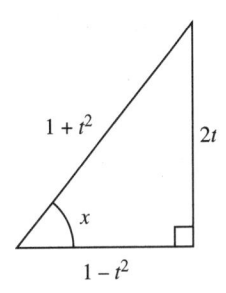

Fig. 7-I/50

and by applying Osborne's rule, that is, replace $\sin^2 x$ by $-\sinh^2 x$, or better still, replace $\sin x$ by $i \sinh x$.

Hyperbolic Functions

WORKED EXAMPLE 114

Determine the following integrals by making appropriate t-formulae substitutions.

(i) $\displaystyle\int \operatorname{sech} x \, dx$

(ii) $\displaystyle\int \operatorname{cosech} x \, dx$

(iii) $\displaystyle\int \dfrac{dx}{1 - \sinh x}$

(iv) $\displaystyle\int \dfrac{dx}{1 + \cosh x}$

(v) $\displaystyle\int \dfrac{dx}{2 + \sinh x}$

(vi) $\displaystyle\int \dfrac{\sinh x}{\cosh x + \sinh x} \, dx.$

Solution 114

(i) $\int \text{sech}\, x \, dx = \int \dfrac{1-t^2}{1+t^2} \dfrac{2dt}{(1-t^2)}$

$= 2\int \dfrac{dt}{1+t^2} = 2\tan^{-1} t$

$= 2\tan^{-1} \tanh \dfrac{x}{2} + c.$

(ii) $\int \text{cosech}\, x \, dx = \int \dfrac{1-t^2}{2t} \dfrac{2\,dt}{(1-t^2)}$

$= \int \dfrac{dt}{t} = \ln t$

$= \ln \tanh \dfrac{x}{2} + c.$

(iii) $\int \dfrac{dx}{1 - \sinh x}$

$= \int \dfrac{\frac{2}{1-t^2}}{1 - \frac{2t}{1-t^2}} \, dt$

$= \int \dfrac{2dt}{1 - t^2 - 2t}$

$= -\int \dfrac{2dt}{t^2 + 2t - 1}$

$= -\int \dfrac{2dt}{(t+1)^2 - 2}$

$= -2 \int \dfrac{dt}{\left(t + 1 - \sqrt{2}\right)\left(t + 1 + \sqrt{2}\right)}$

$\dfrac{1}{\left(t + 1 - \sqrt{2}\right)\left(t + 1 + \sqrt{2}\right)}$

$\equiv \dfrac{A}{t + 1 - \sqrt{2}} + \dfrac{B}{t + 1 - \sqrt{2}}$

$1 \equiv A\left(t + 1 + \sqrt{2}\right) + B\left(t + 1 - \sqrt{2}\right)$

If $t + 1 = -\sqrt{2}$, $B = -\dfrac{1}{2\sqrt{2}}$;

if $t + 1 = \sqrt{2}$,

$A = \dfrac{1}{2\sqrt{2}}$;

$\dfrac{1}{\left(t + 1 - \sqrt{2}\right)\left(t + 1 + \sqrt{2}\right)}$

$\equiv \dfrac{1}{2\sqrt{2}\left(t + 1 - \sqrt{2}\right)}$

$- \dfrac{1}{2\sqrt{2}\left(t + 1 + \sqrt{2}\right)}$

$-2 \int \dfrac{dt}{\left(t + 1 - \sqrt{2}\right)\left(t + 1 + \sqrt{2}\right)}$

$= \int \dfrac{1}{\sqrt{2}\left(t + 1 + \sqrt{2}\right)} \, dt$

$- \int \dfrac{1}{\sqrt{2}\left(t + 1 - \sqrt{2}\right)} \, dt$

$= \dfrac{1}{\sqrt{2}} \ln \left(t + 1 + \sqrt{2}\right)$

$- \dfrac{1}{\sqrt{2}} \ln \left(t + 1 - \sqrt{2}\right) + \ln A$

$= \ln A \dfrac{\left(t + 1 + \sqrt{2}\right)^{\frac{1}{\sqrt{2}}}}{\left(t + 1 - \sqrt{2}\right)^{\frac{1}{\sqrt{2}}}}$

$= \ln A \dfrac{\left(\tanh \frac{x}{2} + 1 + \sqrt{2}\right)^{\frac{1}{\sqrt{2}}}}{\left(\tanh \frac{x}{2} + 1 - \sqrt{2}\right)^{\frac{1}{\sqrt{2}}}}.$

(iv) $\int \dfrac{dx}{1 + \cosh x} = \int \dfrac{\frac{2}{1-t^2}}{1 + \frac{1+t^2}{1-t^2}} \, dt$

$= \int \dfrac{2dt}{1 - t^2 + 1 + t^2}$

$= \int dt$

$= \tanh \dfrac{x}{2} + c.$

(v) $\displaystyle\int \frac{dx}{2+\sinh x} = \int \frac{\frac{2}{(1-t^2)}}{2+\frac{2t}{(1-t^2)}} dt$

$\displaystyle = \int \frac{dt}{1-t^2+t}$

$\displaystyle = -\int \frac{dt}{t^2-t-1}$

$\displaystyle = -\int \frac{dt}{\left(t-\frac{1}{2}\right)^2 - \frac{1}{4} - 1}$

$\displaystyle = -\int \frac{dt}{\left(t-\frac{1}{2}\right)^2 - \left(\frac{\sqrt{5}}{2}\right)^2}$

$\displaystyle = \int \frac{dt}{\left(\frac{\sqrt{5}}{2}\right)^2 - \left(t-\frac{1}{2}\right)^2}$

$\displaystyle \frac{1}{\left[\frac{\sqrt{5}}{2} - \left(t-\frac{1}{2}\right)\right]\left[\frac{\sqrt{5}}{2} + \left(t-\frac{1}{2}\right)\right]}$

$\displaystyle \equiv \frac{A}{\frac{\sqrt{5}}{2} - \left(t-\frac{1}{2}\right)} + \frac{B}{\frac{\sqrt{5}}{2} + \left(t-\frac{1}{2}\right)}$

$\displaystyle 1 \equiv A\left[\frac{\sqrt{5}}{2} + \left(t-\frac{1}{2}\right)\right] + B\left[\frac{\sqrt{5}}{2} - \left(t-\frac{1}{2}\right)\right].$

If $t - \frac{1}{2} = \frac{\sqrt{5}}{2}$, $A = \frac{1}{\sqrt{5}}$;

if $t - \frac{1}{2} = -\frac{\sqrt{5}}{2}$, $B = \frac{1}{\sqrt{5}}$

$\displaystyle \int \frac{dt}{\left(\frac{\sqrt{5}}{2}\right)^2 - \left(t-\frac{1}{2}\right)^2}$

$\displaystyle = \int \frac{\frac{1}{\sqrt{5}}}{\frac{\sqrt{5}}{2} - \left(t-\frac{1}{2}\right)} dt$

$\displaystyle + \int \frac{\frac{1}{\sqrt{5}}}{\frac{\sqrt{5}}{2} + \left(t-\frac{1}{2}\right)} dt$

$\displaystyle = -\frac{1}{\sqrt{5}} \ln\left|\left[\frac{\sqrt{5}}{2} - \left(t-\frac{1}{2}\right)\right]\right|$

$\displaystyle + \frac{1}{\sqrt{5}} \ln\left|\frac{\sqrt{5}}{2} + \left(t-\frac{1}{2}\right)\right| + \ln A$

$\displaystyle = \ln A \left(\frac{\frac{\sqrt{5}}{2} + \left(t-\frac{1}{2}\right)}{\frac{\sqrt{5}}{2} - \left(t-\frac{1}{2}\right)}\right)^{\frac{1}{\sqrt{5}}}$

(vi) $\displaystyle\int \frac{\sinh x}{\cosh x + \sinh x} dx = \int \frac{\frac{2t}{1-t^2}}{\frac{1+t^2}{1-t^2} + \frac{2t}{1-t^2}} \frac{2\,dt}{(1-t^2)}$

$\displaystyle = \int \frac{\frac{4t}{1-t^2}}{1+2t+t^2} dt$

$\displaystyle = \int \frac{4t\,dt}{(t+1)^2(1-t^2)}$

$\displaystyle = \int \frac{4t\,dt}{(1-t)(1+t)^3}$

$\displaystyle \frac{4t}{(1-t)(1+t)^3} \equiv \frac{A}{1-t} + \frac{B}{1+t}$

$\displaystyle + \frac{C}{(1+t)^2} + \frac{D}{(1+t)^3}$

$4t = A(1+t)^3 + B(1+t)^2(1-t)$
$\qquad + C(1+t)(1-t) + D(1-t)$

If $t = 1$

$A = \dfrac{4}{8} = \dfrac{1}{2}$

$\boxed{A = \dfrac{1}{2}}$

If $t = 0$

$A + B + C + D = 0$

$B + C = -A - D$

$= -\dfrac{1}{2} + 2$

$B + C = \dfrac{3}{2}$ ...(1)

If $t = -1$

$D = -\dfrac{4}{2} = -2$

$\boxed{D = -2}$

If $t = 2$

$27A - 9B - 3C - D = 8$

$\dfrac{27}{2} - 9B - 3C + 2 = 8$

$9B + 3C = -8 + 2 + \dfrac{27}{2}$

$= \dfrac{15}{2}$

$9B + 3C = \dfrac{15}{2}$ ...(2)

$\begin{array}{ll} B + C = \dfrac{3}{2} & 3B + 3C = \dfrac{9}{2} \\ 9B + 3C = \dfrac{15}{2} & 9B + 3C = \dfrac{15}{2} \end{array}$

$6B = \dfrac{15}{2} - \dfrac{9}{2} = \dfrac{6}{2} = 3$

$\boxed{B = \dfrac{1}{2}}$

$\boxed{C = 1}$

$\dfrac{4t}{(1-t)(1+t)^3} \equiv \dfrac{\tfrac{1}{2}}{1-t} + \dfrac{\tfrac{1}{2}}{1+t}$

$\qquad\qquad + \dfrac{1}{(1+t)^2} - \dfrac{2}{(1+t)^3}$

$\displaystyle\int \dfrac{4t\, dt}{(1-t)(1+t)^3}$

$\displaystyle\equiv \int \dfrac{\tfrac{1}{2}}{1-t}dt + \int \dfrac{\tfrac{1}{2}}{1+t}dt$

$\displaystyle + \int \dfrac{1}{(1+t)^2}dt - \int \dfrac{2\, dt}{(1+t)^3}$

$= -\dfrac{1}{2}\ln(1-t) + \dfrac{1}{2}\ln(1+t)$

$\qquad - \dfrac{1}{(1+t)} + \dfrac{1}{(1+t)^2} + c$

$= -\dfrac{1}{2}\ln\left(1 - \tanh\dfrac{x}{2}\right)$

$\qquad + \dfrac{1}{2}\ln\left(1 + \tanh\dfrac{x}{2}\right)$

$\qquad - \dfrac{1}{\left(1 + \tanh\dfrac{x}{2}\right)}$

$\qquad + \dfrac{1}{\left(1 + \tanh\dfrac{x}{2}\right)^2} + c$

$= \ln \dfrac{\left(1 + \tanh\dfrac{x}{2}\right)^{\tfrac{1}{2}}}{\left(1 - \tanh\dfrac{x}{2}\right)^{\tfrac{1}{2}}}$

$\qquad - \dfrac{1}{\left(1 + \tanh\dfrac{x}{2}\right)}$

$\qquad + \dfrac{1}{\left(1 + \tanh\dfrac{x}{2}\right)^2} + c.$

Note that most of the above could have been solved by using the basic definitions of sinh x and cosh x.

Circular Functions

$$\int \sec x \, dx = \int \frac{1+t^2}{1-t^2} \cdot \frac{2dt}{1+t^2}$$

$$= 2 \int \frac{1}{1-t^2} \, dt$$

$$= 2 \int \frac{1}{(1-t)(1+t)} \, dt$$

$$= \int \frac{1}{1-t} dt + \int \frac{1}{1+t} dt$$

$$= -\ln(1-t) + \ln(1+t)$$

$$= \ln \frac{1+t}{1-t}$$

$t = \tan \frac{x}{2}$

$$\frac{dt}{dx} = \frac{1}{2} \sec^2 \frac{x}{2}$$

$$\frac{dt}{dx} = \frac{1}{2}\left(1 + \tan^2 \frac{x}{2}\right)$$

$$dx = \frac{2 \, dt}{1+t^2}$$

$$\boxed{\int \sec x \, dx = \ln\left(\frac{1 + \tan \frac{x}{2}}{1 - \tan \frac{x}{2}}\right)}$$

$$\frac{1}{(1-t)(1+t)} \equiv \frac{A}{1-t} + \frac{B}{1+t}$$

$$1 \equiv A(1+t) + B(1-t)$$

if $t = 1$, $A = \frac{1}{2}$

if $t = -1$, $B = \frac{1}{2}$

but $\sec x + \tan x = \frac{1+t^2}{1-t^2} + \frac{2t}{1-t^2}$ from the triangle above

$$= \frac{(1+t)^2}{(1-t)^2} = \frac{(1+t)}{(1-t)} = \frac{1 + \tan \frac{x}{2}}{1 - \tan \frac{x}{2}}$$

$$\boxed{\sec x \, dx = \ln(\sec x + \tan x) + c}$$

Fig. 7-I/51

$$\int \csc x \, dx = \int \frac{1+t^2}{2t} \cdot \frac{2 \, dt}{1+t^2}$$

$$= \int \frac{1}{t} dt = \ln \tan \frac{x}{2}$$

$$\cot x + \csc x = \frac{1-t^2}{2t} + \frac{1+t^2}{2t}$$

$$= \frac{1-t^2+1+t^2}{2t}$$

$$= \frac{1}{t} = \frac{1}{\tan \frac{x}{2}}$$

$$\tan \frac{x}{2} = \frac{1}{\cot x + \csc x}$$

$$\int \csc x \, dx = \ln \left| \tan \frac{x}{2} \right|$$

$$= \ln \left| \frac{1}{\cot x + \csc x} \right|$$

$$\boxed{\int \csc x \, dx = -\ln|\cot x + \csc x| + c}$$

$$\int \frac{dx}{a \sin x + b \cos x}$$

$a \sin x + b \cos x = R \sin(x + \alpha)$

$a \sin x + b \cos x = R \sin x \cos \alpha + R \sin \alpha \cos x$

$a = R \cos \alpha, \quad b = R \sin \alpha$

$R = \sqrt{a^2 + b^2}$

$\tan \alpha = \dfrac{b}{a} \Rightarrow \alpha = \tan^{-1} \dfrac{b}{a}$

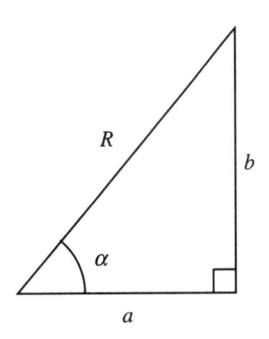

Fig. 7-I/52

$$\int \frac{dx}{a\sin x + b\cos x}$$

$$= \int \frac{dx}{\sqrt{a^2+b^2}\sin(x-\alpha)}$$

$$= \frac{1}{\sqrt{a^2+b^2}} \int \operatorname{cosec}(x-\alpha)\,dx$$

$$= \frac{1}{\sqrt{a^2+b^2}}$$

$$(-\ln|\cot(x-\alpha)+\operatorname{cosec}(x-\alpha)|)+c$$

WORKED EXAMPLE 115

Determine the indefinite integrals:-

(i) $\displaystyle\int \frac{d\theta}{3\sin\theta + 4\cos\theta}$

(ii) $\displaystyle\int \frac{d\theta}{\sin\theta - \cos\theta}$

Solution 115

(i) $\displaystyle\int \frac{d\theta}{3\sin\theta + 4\cos\theta}$

$$= \int \frac{\frac{2}{1+t^2}}{3\times\frac{2t}{1+t^2}+4\frac{(1-t^2)}{1+t^2}}\,dt$$

$$= \int \frac{2\,dt}{6t+4-4t^2}$$

$$= \int \frac{dt}{-2t^2+3t+2}$$

Fig 7-I/53 (triangle with sides $1+t^2$, $2t$, $1-t^2$, angle x)

$t = \tan\frac{\theta}{2}$

$\dfrac{dt}{d\theta} = \dfrac{1}{2}\left(1+\tan^2\dfrac{\theta}{2}\right)$

$d\theta = \dfrac{2\,dt}{1+t^2}$

$$\int \frac{dt}{-2\left(t^2-\frac{3}{2}t-1\right)}$$

$$= -\frac{1}{2}\int \frac{dt}{\left(t-\frac{3}{4}\right)^2 - \frac{9}{16}-1}$$

$$= -\frac{1}{2}\int \frac{dt}{\left(t-\frac{3}{4}\right)^2 - \left(\frac{5}{4}\right)^2}$$

$$= -\frac{1}{2}\cdot\frac{4}{2(5)} \ln \frac{\left(t-\frac{3}{4}-\frac{5}{4}\right)}{\left(t-\frac{3}{4}+\frac{5}{4}\right)}$$

$$= -\frac{1}{5}\ln \frac{(t-2)}{\left(t+\frac{1}{2}\right)}$$

$$= -0.2\ln\left|\frac{\tan\frac{\theta}{2}-2}{\tan\frac{\theta}{2}+\frac{1}{2}}\right|$$

where

$$\int \frac{dx}{x^2-a^2} = \frac{1}{2a}\int\left(\frac{1}{x-a}-\frac{1}{x+a}\right)dx$$

$$= \frac{1}{2a}\ln(x-a) - \frac{1}{2a}\ln(x+a)$$

$$= \frac{1}{2a}\ln\frac{x-a}{x+a}$$

$\dfrac{1}{x^2-a^2} \equiv \dfrac{A}{x-a} + \dfrac{B}{x+a}$

or $1 \equiv A(x+a) + B(x-a)$

if $x=a$, $A = \dfrac{1}{2}a$;

if $x=-a$, $B = -\dfrac{1}{2a}$

$\dfrac{1}{x^2-a^2} \equiv \dfrac{1}{2a(x-a)}$

$\qquad -\dfrac{1}{2a(x+a)}.$

Alternatively

$$\int \frac{d\theta}{3\sin\theta + 4\cos\theta}$$

$3\sin\theta + 4\cos\theta = R\sin(\theta + \alpha)$
$\equiv R\sin\theta\cos\alpha + R\sin\alpha\cos\theta$

$3 = R\cos\alpha,\ 4 = R\sin\alpha$

$R = 5,\ \alpha = \tan^{-1}\dfrac{4}{3} = 53.13°$

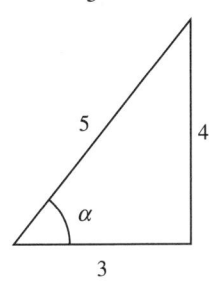

Fig. 7-I/54

$$\int \frac{d\theta}{5\sin(\theta + 53.13°)}$$

$$= 0.2\int \frac{d\theta}{\sin(\theta + 53.13°)}$$

$$= 0.2\int \mathrm{cosec}\,(\theta + 53.13°)\,d\theta$$

$$= -0.2\ln\left|\cot(\theta + 53.13°) + \mathrm{cosec}\,(\theta + 53.13°) + c\right|$$

(ii) $\displaystyle\int \frac{d\theta}{\sin\theta - \cos\theta}$

$$= \int \frac{d\theta}{\sqrt{2}\sin\left(\theta - \dfrac{\pi}{4}\right)}$$

$$= \int 0.707\,\mathrm{cosec}\left(\theta - \frac{\pi}{4}\right)d\theta$$

$$= -0.707\ln\left|\cot\left(\theta - \frac{\pi}{4}\right) + \mathrm{cosec}\left(\theta - \frac{\pi}{4}\right)\right|$$

$\sin\theta - \cos\theta = R\sin(\theta - \alpha)$
$= R\sin\theta\cos\alpha - R\cos\theta\sin\alpha$

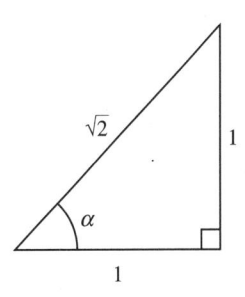

$1 = R\cos\alpha,\qquad 1 = R\sin\alpha$

Fig. 7-I/55

Exercises 16

1. Write down

    (i) $\displaystyle\int \sec x\,dx$

    (ii) $\displaystyle\int \mathrm{cosec}\,x\,dx$ in terms of single angles, x.

2. Write down

    (i) $\displaystyle\int \sec x\,dx$

    (ii) $\displaystyle\int \mathrm{cosec}\,x\,dx$ in terms of half angles, $\dfrac{x}{2}$.

3. Prove the integrals in the questions 1 and 2.

4. (i) $\displaystyle\int \frac{d\theta}{7\cos\theta + 5\sin\theta}$

    (ii) $\displaystyle\int \frac{dx}{\sin\left(x + \dfrac{\pi}{4}\right)}$

    (iii) $\displaystyle\int \frac{dx}{\cos\left(x - \dfrac{\pi}{3}\right)}$.

5. $\displaystyle\int \frac{d\theta}{a\cos\theta + b\sin\theta}$.

6. (i) $\displaystyle\int (5\sin\theta + 2\cos\theta)^{-1}\,d\theta$

    (ii) $\displaystyle\int \frac{1}{15\sin\theta + 8\cos\theta}\,d\theta$.

7. (i) $\displaystyle\int \frac{\sin x}{1 - \cos x}\, dx$

   (ii) $\displaystyle\int \frac{\cos x}{\sin x + 1}\, dx$.

8. (i) $\displaystyle\int \csc\left(3x - \frac{\pi}{3}\right) dx$

   (ii) $\displaystyle\int \sec\left(2x + \frac{\pi}{4}\right) dx$.

9. By substituting $\sin x = 2 \sin \dfrac{x}{2} \cos \dfrac{x}{2}$, show that the integral
$$\int \csc dx = \ln \tan \frac{x}{2} + c.$$

10. By substituting $\cos x = \sin\left(x + \dfrac{\pi}{2}\right)$ show that the integral
$$\int \sec x \, dx = \ln \tan \frac{1}{2}\left(x + \frac{\pi}{2}\right) + c$$
$$= \ln \left(\frac{1 + \tan \frac{x}{2}}{1 - \tan \frac{x}{2}}\right) + c.$$

17

The Mean or Average Values of Functions Over a Given Range

The rectangle shown in Fig. 7-I/56 has a base, b and a height, h.

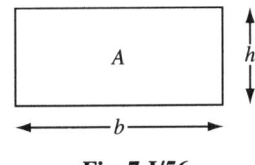

Fig. 7-I/56

The mean value over the base is obviously, h, the height is uniform throughout this base length.

If the area A is known and the base width is b, then the mean value or height may be found.

$$A = bh, \quad h = \frac{A}{b}$$

by dividing the area, A, by the base width.

In order to determine the mean value of a half-wave of a sinusoidal waveform, we divide the area of the half sinewave by the base, π.

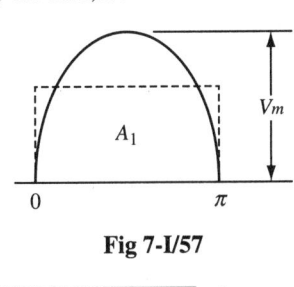

Fig 7-I/57

Fig 7-I/58

$$\text{Mean value} = \frac{\text{area of the half sinewave}}{\text{the base}}.$$

The area of the half sinewave can be found by integration between the limits 0 and π.

$$\int_0^\pi V_m \sin\theta \, d\theta$$

where the instantaneous value of the sinewave is given by $v = V_m \sin\theta$ and V_m is the peak or maximum value of the waveform.

$$\text{Mean value} = \frac{\int_0^\pi V_m \sin\theta \, d\theta}{\pi} = \frac{V_m(-\cos\theta)\big|_0^\pi}{\pi}$$

$$= \frac{V_m}{\pi}[-\cos\pi - (-\cos 0)]$$

$$= \frac{V_m}{\pi}[-(-1) + 1] = \frac{2V_m}{\pi} = 0.637\, V_m.$$

The mean value or the average value of the voltage of a sinewave over half a cycle is $0.637\, V_m$.

The mean value or the average value of the voltage or current of a sinewave over a complete cycle is 0 since the positive area cancels the negative area.

$$\text{Mean value} = \frac{\int_0^{2\pi} V_m \sin\theta \, d\theta}{2\pi} = \frac{V_m(-\cos\theta)\big|_0^{2\pi}}{2\pi}$$

$$= \frac{V_m}{2\pi}[-\cos 2\pi - (-\cos 0)]$$

$$= \frac{V_m}{2\pi}(-1 + 1) = 0.$$

Worked Example 116

Find the mean values of the waveforms whose shapes for one cycle are as shown:-

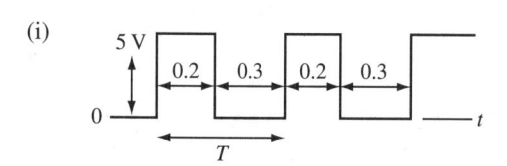

(i) [5 V square wave with widths 0.2, 0.3, 0.2, 0.3 over period T]

Fig. 7-I/59

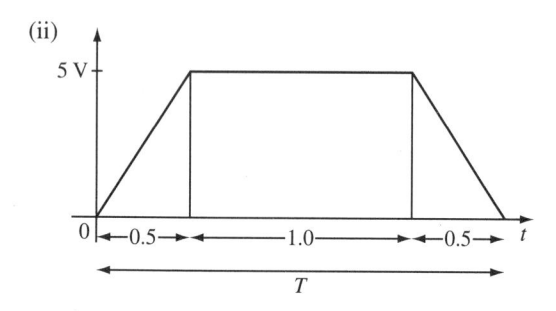

(ii) [5 V trapezoidal waveform with ramp 0.5, flat 1.0, ramp 0.5 over period T]

Fig. 7-I/60

Solution 116

(i) The area over one cycle $= 5 \times 0.2 = 1.0$.

$$\text{The mean value} = \frac{\text{The area over one cycle}}{\text{The time taken}}$$

$$= \frac{1.0}{0.2 + 0.3}$$

$$= 2\,V.$$

(ii) The area of one complete cycle

$$= \frac{1}{2}(1.0 + 2.0)\,5$$

$$= 7.5$$

$$= \text{the area of the trapezium.}$$

$$\text{The mean value} = \frac{7.5}{2} = 3.75\,V.$$

Root Mean Square (R.M.S.)

To determine and define the Root Mean Square value of a sinewave.

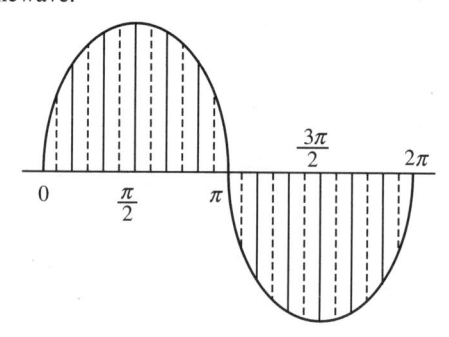

Fig. 7-I/61

Divide the base into twelve equal intervals of width $\frac{\pi}{6}$, the mean height between 0 and $\frac{\pi}{6}$ is $\frac{\pi}{12}$.

The R.M.S. value of $i = I_m \sin \omega t$ over one complete cycle can be found by integration.

$$\text{Mean Squares} = \frac{1}{2\pi} \int_0^{2\pi} i^2 \, d(\omega t)$$

$$\text{Root Mean Squares} = \sqrt{\frac{1}{2\pi} \int_0^{2\pi} i^2 \, d(\omega t)}$$

but $i = I_m \sin \omega t$

$$\int_0^{2\pi} i^2 d(\omega t) = \int_0^{2\pi} I_m^2 \sin^2 \omega t \, d(\omega t)$$

to evaluate this definite integral we need to substitute $\sin^2 \omega t$ in terms of the double angle $\cos 2\omega t$.

$\cos(A + B) = \cos A \cos B - \sin A \sin B$, if $B = A$

$\cos(A + A) = \cos A \cos A - \sin A \sin A$

$\cos 2A = \cos^2 A - \sin^2 A$

but $\sin^2 A + \cos^2 A = 1 \Rightarrow 1 - \sin^2 A = \cos^2 A$

$\cos 2A = 1 - \sin^2 A - \sin^2 A$

$\cos 2A = 1 - 2\sin^2 A$

$2\sin^2 A = 1 - \cos 2A$

$\sin^2 A = \dfrac{1 - \cos 2A}{2}$

therefore $\sin^2 \omega t = \dfrac{1 - \cos 2\omega t}{2}$

The Mean or Average Values of Functions Over a Given Range

$$\frac{I_m^2}{2} \int_0^{2\pi} (1 - \cos 2\omega t)\, d(\omega t)$$

$$= \frac{I_m^2}{2} \left(\omega t - \frac{\sin 2\omega t}{2} \right)_0^{2\pi}$$

$$= \frac{I_m^2}{2} \left(2\pi - \frac{\sin 4\pi}{2} - 0 \right) = \pi I_m^2.$$

Let $i = I_m \sin \omega t$ where i is the instantaneous value of the current, I_m is the peak value, ω is the angular velocity and t is the time. Note that i and t are the variables.

ωt		$\sin \omega t$	
$\frac{\pi}{12}$	15°	0.259	i_1
$\frac{3\pi}{12}$	45°	0.707	i_2
$\frac{5\pi}{12}$	75°	0.966	i_3
$\frac{7\pi}{12}$	105°	0.966	i_4
$\frac{9\pi}{12}$	135°	0.707	i_5
$\frac{11\pi}{12}$	165°	0.259	i_6
$\frac{13\pi}{12}$	195°	−0.259	i_7
$\frac{15\pi}{12}$	225°	−0.707	i_8
$\frac{17\pi}{12}$	255°	−0.966	i_9
$\frac{19\pi}{12}$	285°	−0.966	i_{10}
$\frac{21\pi}{12}$	315°	−0.707	i_{11}
$\frac{23\pi}{12}$	345°	−0.259	i_{12}

Mean squares

$$= \frac{\left(i_1^2 + i_2^2 + i_3^2 + i_4^2 + i_5^2 + i_6^2 + i_7^2 + i_8^2 + i_9^2 + i_{10}^2 + i_{11}^2 + i_{12}^2 \right)}{12}$$

$$= \left(0.259^2 + 0.707^2 + 0.966^2 \right.$$
$$+ 0.966^2 + 0.707^2 + 0.259^2$$
$$+ (-0.259)^2 + (-0.707)^2 + (-0.966)^2$$
$$\left. + (-0.966)^2 + (-0.707)^2 + (-0.259)^2 \right) \frac{I_m^2}{12}$$

$$= \frac{4 \times 0.259^2 + 4 \times 0.707^2 + 4 \times 0.977^2}{12} I_m^2$$

$$= \frac{0.259^2 + 0.707^2 + 0.966^2}{3} I_m^2 = 0.5 I_m^2.$$

Mean squares $= \frac{1}{2} I_m^2$

R.M.S. $= I_m \sqrt{\frac{1}{2}} = \frac{I_m}{\sqrt{2}}$

$$\boxed{I = \frac{I_m}{\sqrt{2}}}$$

$$\sqrt{\frac{I_m^2}{2\pi}(\pi)} = \frac{I_m}{\sqrt{2}} = \text{R.M.S.}$$

$$I = \frac{I_m}{\sqrt{2}}.$$

What is the R.M.S. value over a half cycle?

$$\sqrt{\frac{1}{\pi} \int_0^{\pi} i^2 d(\omega t)} = I_m \sqrt{\frac{1}{\pi} \frac{\pi}{2}}$$

$$= \frac{I_m}{\sqrt{2}}$$

$$\frac{1}{2} \int_0^{\pi} (1 - \cos 2\omega t)\, d(\omega t)$$

$$= \frac{1}{2} \left((\omega t) - \frac{\sin 2(\omega t)}{2} \right)_0^{\pi}$$

$$= \frac{1}{2} \pi.$$

Therefore the R.M.S. value of this function over any given range is the same.

WORKED EXAMPLE 117

To find the R.M.S. value of a saw-tooth waveform.

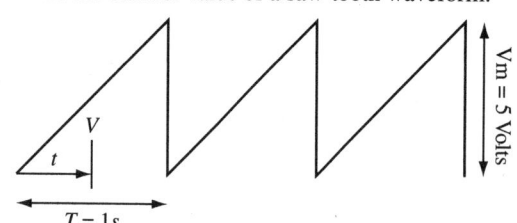

Fig. 7-I/62

Solution 117

At time, t, the voltage is V. $\dfrac{V}{t} = \dfrac{V_m}{T}$

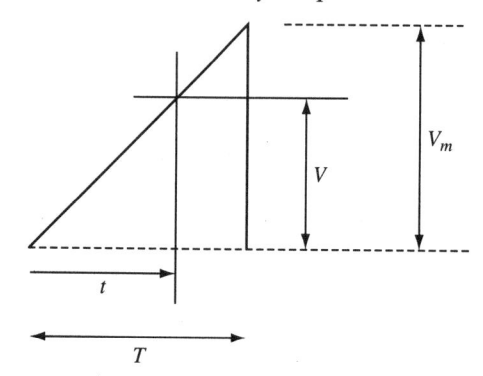

Fig. 7-I/63

$$V = \dfrac{V_m}{T} t$$

$$\text{R.M.S.} = \sqrt{\dfrac{1}{T} \int_0^T V^2 \, dt}$$

$$= \sqrt{\dfrac{1}{T} \int_0^T \left(\dfrac{V_m}{T} t\right)^2 dt}$$

$$\text{R.M.S.} = \sqrt{\dfrac{1}{T} \int_0^T \dfrac{V_m^2 t^2}{T^2} \, dt}$$

$$= \sqrt{\dfrac{1}{T^3} \left(V_m^2 \dfrac{t^3}{3}\right)_0^T} = \sqrt{\dfrac{V_m^2 T^3}{3 T^3}}$$

$$V = \dfrac{V_m}{\sqrt{3}}.$$

WORKED EXAMPLE 118

A current is given by $i = \dfrac{10 \, I_m}{T^2} (Tt - t^2)$ from $t = 0$ to $t = T$. Find an expression for the R.M.S. value from 0 to T. I_m and T are constants.

Solution 118

$$\int_0^T i^2 \, dt = \int_0^T \left(\dfrac{10 \, I_m}{T^2} (Tt - t^2)\right)^2 dt$$

$$= \dfrac{100 \, I_m^2}{T^4} \int_0^T \left(T^2 t^2 - 2Tt^3 + t^4\right) dt$$

$$= \dfrac{100 \, I_m^2}{T^4} \left[\dfrac{1}{3} T^2 t^3 - \dfrac{1}{2} T t^4 + \dfrac{1}{5} t^5\right]_0^T$$

$$= \dfrac{100 \, I_m^2}{T^4} \left[\dfrac{T^5}{3} - \dfrac{1}{2} T^5 + \dfrac{1}{5} T^5\right]$$

$$= \dfrac{100 \, I_m^2 \, T^5}{T^4} \left(\dfrac{10 - 15 + 6}{30}\right)$$

$$= \dfrac{100 \, I_m^2 \, T}{30} = \dfrac{10 \, I_m^2 \, T}{3}$$

$$\text{R.M.S.} = \sqrt{\dfrac{1}{T} \left(\dfrac{10 \, I_m^2 \, T}{3}\right)}$$

$$= I_m \sqrt{\dfrac{10}{3}}$$

$$= \sqrt{\dfrac{1}{T} \int_0^T i^2 \, dt}$$

$$= \sqrt{\dfrac{1}{T} \dfrac{10}{3} I_m^2 \, T}$$

$$= I_m \sqrt{\dfrac{10}{3}}.$$

Exercises 17

1. Determine the mean values of the following:-

   (i) $y = \sin 2x$ between $x = 0$ and $x = \dfrac{\pi}{2}$.

   (ii) $y = 2\sin \dfrac{1}{2} x$ between $x = 0$ and $x = \pi$.

   (iii) $y = \sin 3t$ between $t = 0$ and $t = \dfrac{\pi}{2}$.

   (iv) $y = 1 + 2\sin\theta$ between $\theta = 0$ and $\theta = \pi$.

   (v) $y = e^{2x}$ between $x = 0$ and $x = 2$.

2. If $v = 200 \sin \omega t$, $i = 200 \sin\left(\omega t + \dfrac{\pi}{6}\right)$, find the mean values for each quantity from $t = 0$ to $\dfrac{\pi}{\omega}$.

3. If $v = 3 + 4\cos \omega t$, find the r.m.s. value from $t = 0$ to $\dfrac{2\pi}{\omega}$.

4. Evaluate

   (i) $\displaystyle\int_0^{\frac{\pi}{3}} \cos\left(2x - \dfrac{\pi}{6}\right) dx$

   (ii) $\displaystyle\int_1^2 \left(e^{3x} - e^{-3x}\right) dx$

   (iii) $\displaystyle\int_0^{\frac{\pi}{2}} \left[1 - \sin\left(2x - \dfrac{\pi}{.6}\right)\right] dx$.

5. Find the mean value of $\cos^2 \omega t$ between $t = 0$ and $t = \dfrac{2\pi}{\omega}$.

6. Find the r.m.s value of $v = V_m \sin \omega t$ over the range $t = 0$ to $t = \dfrac{2\pi}{\omega}$.

General Standard Integrals

Prove the following standard integrals.

$$\int ax^n \, dx = \frac{ax^{n+1}}{n+1} + c$$

where $n \neq -1$

$$\int (ax+b)^n \, dx = \frac{(ax+b)^{n+1}}{a(n+1)} + c$$

Where $n \neq -1$

$$\int \frac{1}{x} \, dx = \ln x + c$$

$$\int \frac{1}{(ax+b)} \, dx = \frac{1}{a} \ln |(ax+b)| + c$$

$$\int \frac{dx}{a^2 x^2 + b^2} = \frac{1}{ab} \tan^{-1} \frac{ax}{b} + c$$

$$\int \frac{x}{x^2 + a^2} \, dx = \frac{1}{2} \ln \left| A(x^2 + a^2) \right|$$

$$\int \frac{dx}{x^2 + a^2} = \frac{1}{a} \tan^{-1} \frac{x}{a} + c$$

$$\int \frac{dx}{x(x^2 + a^2)} = \frac{1}{2a^2} \ln \frac{x^2}{x^2 + a^2} + c$$

$$\int \frac{dx}{a^2 x^2 - b^2} = \frac{1}{2ab} \ln \left| A \frac{ax-b}{ax+b} \right| + c$$

$$\int \frac{x}{x^2 - a^2} \, dx = \frac{1}{2} \ln A \left| (x^2 - a^2) \right| + c$$

$$\int \frac{dx}{x \pm 1} = \ln |A(x \pm 1)| + c$$

$$\int \frac{x^2}{\sqrt{x^2 + a^2}} \, dx = \frac{x}{2} \sqrt{x^2 + a^2}$$
$$- \frac{a^2}{2} \ln A \left(x + \sqrt{x^2 + a^2} \right) + c$$

$$\int \frac{dx}{x^2 \sqrt{x^2 + a^2}} = -\frac{1}{a^2} \frac{\sqrt{x^2 + a^2}}{x} + c$$

$$\int x \sqrt{x^2 + a^2} \, dx = \frac{1}{3} \left(x^2 + a^2 \right)^{\frac{3}{2}} + c$$

$$\int x^2 \sqrt{x^2 + a^2} \, dx = \frac{x}{8} \left(2x^2 + a^2 \right) \sqrt{x^2 + a^2}$$
$$- \frac{a^2}{8} \ln A \left(x + \sqrt{x^2 + a^2} \right) + c$$

$$\int \frac{\sqrt{x^2 + a^2}}{x} \, dx = \sqrt{x^2 + a^2} + \frac{a}{2} \ln A \frac{\sqrt{x^2 + a^2} - a}{\sqrt{x^2 + a^2} + a}$$

$$\int \frac{dx}{x \sqrt{x^2 - a^2}} = -\frac{1}{a} \sin^{-1} \frac{a}{|x|} + c$$

$$= \tan^{-1} \frac{x}{\sqrt{a^2 - x^2}} + c$$

$$\int \frac{x}{\sqrt{a^2 - x^2}} \, dx = -\sqrt{a^2 - x^2} + c$$

$$\int e^{kx} \, dx = \frac{1}{k} e^{kx} + c$$

$$\int a^x \, dx = \frac{a^x}{\ln a} + c$$

$$\int x e^{-x^2} \, dx = -\frac{1}{2} e^{-x^2} + c$$

$$\int \ln x \, dx = x \ln x - x + c$$

$$\int \sin x \, dx = -\cos x + c$$

$$\int \cos x \, dx = \sin x + c$$

$$\int \sin^2 x \, dx = -\frac{1}{2} \sin x \cos x + \frac{x}{2} + c$$

$$\int \cosec x \, dx = \ln \left| \tan \frac{x}{2} \right| + c$$

$$\int \operatorname{cosec} x \, dx = \ln \tan \frac{x}{2} + c$$
$$= -\ln |\operatorname{cosec} x + \cot x| + c$$

$$\int \operatorname{cosec}^2 x \, dx = -\cot x + c$$

$$\int \cos^2 x \, dx = \frac{1}{2} \sin x \cos x + \frac{x}{2} + c$$

$$\int \sec^2 dx = \tan x + c$$

$$\int \tan x \, dx = -\ln \cos x + c$$

$$\int \tan^2 x \, dx = \tan x - x + c$$

$$\int \cot x \, dx = \ln |\sin x| + c$$

$$\int \cot^2 x \, dx = -\cot x - x + c$$

$$\int \sinh x \, dx = \cosh x + c$$

$$\int \cosh x \, dx = \sinh x + c$$

$$\int \frac{dx}{\sinh x} = \ln \tanh \frac{x}{2} + c = \ln \frac{\cosh x - 1}{\sinh x} + c$$

$$\int \operatorname{cosech}^2 x \, dx = -\coth x + c$$

$$\int \operatorname{sech} x \, dx = \sin^{-1}(\tanh x) + c = \cos^{-1} \frac{1}{\cosh x} + c$$
$$= 2 \tan^{-1}\left(\tanh \frac{x}{2}\right) + c$$

$$\int \operatorname{sech}^2 x \, dx = \tanh x + c$$

$$\int \sinh^{-1} \frac{x}{a} \, dx = x \sinh^{-1} \frac{x}{a} - \sqrt{x^2 + a^2} + c$$

$$\int \cosh^{-1} \frac{x}{a} \, dx = x \cosh^{-1} \frac{x}{a} - \sqrt{x^2 - a^2} + c$$

$$\int \sec x \, dx = \ln |\sec x + \tan x| + c$$
$$= \ln \left| \frac{1 + \tan \frac{x}{2}}{1 - \tan \frac{x}{2}} \right|$$

$$\int x \sinh^{-1} \frac{x}{a} \, dx = \frac{2x^2 + a^2}{4} \sinh^{-1} \frac{x}{a} - \frac{x}{4} \sqrt{x^2 + a^2} + c$$

$$\int x \cosh^{-1} \frac{x}{a} \, dx = \frac{2x^2 - a^2}{4} \cosh^{-1} \frac{x}{a} - \frac{x}{4} \sqrt{x^2 - a^2} + c$$

$$\int_0^a \frac{dx}{\sqrt{x^2 + a^2}} = \ln\left(1 + \sqrt{2}\right)$$

$$\int_0^a \frac{x \, dx}{\sqrt{x^2 + a^2}} = a\left(\sqrt{2} - 1\right)$$

$$\int_0^a \frac{x^2 \, dx}{\sqrt{x^2 + a^2}} = \frac{a^2}{2}\left(\sqrt{2} - \ln\left(1 + \sqrt{2}\right)\right)$$

$$\int_0^1 \frac{\ln x}{x - 1} \, dx = \frac{\pi^2}{6}$$

$$\int_0^1 \frac{\ln x}{x + 1} \, dx = -\frac{\pi^2}{12}$$

$$\int_0^a \tanh^{-1} \frac{x}{a} \, dx = a \ln 2$$

$$\int_0^a \sin^{-1} \frac{x}{a} \, dx = a\left(\frac{\pi}{2} - 1\right)$$

$$\int_0^a \cos^{-1} \frac{x}{a} \, dx = a$$

$$\int_0^1 \sin^{-1} x \, \frac{dx}{x} = \frac{\pi}{2} \ln 2.$$

7. INTEGRAL CALCULUS APPLICATIONS

Index

Acceleration 6

Algebraic function 2

Approximate
    numerical integration 83

Arbitrary constant calculation 1

Arc in polar
    coordinates length of 92

Area
    under the curve 1, 9–12
    of surface of revolution 97–99

Astroid 98–99

Auxiliary equation 84–85

Centre of gravity 72

Centroid
    of an area 72–75
    of a triangle 75
    of a volume 72–73
    of a solid of 75–76
    revolution 72

Change of variable 7

Circular functions graph of 15

Complementary functions 86–7

Constant of integration 1–2

Definite integration 8

Differential
    equation auxiliary 84
    exact 79
    second order-linear 84
    unequal roots 85
    equal roots 85
    complex roots 85
    integrating factor 80

Displacement 6

Elemental strip 1

Exponential functions integration of 19

First moment of area 74

Formulae of reduction 52

General solution
    of differential equations 88-90

Hyperbolic functions integration of 34

Indefinite integrals
    of hyperbolic functions 34–35

Inverse hyperbolic 32–34

Integral sign 1
Integrals
    algebraic 1–2
    definite 8–9
    hyperbolic 29–31
    indefinite 34–35
    trigonometric 15

Integration
    approximate 63–65
    as the reverse
    of differentiation 1
    by inspection 25
    by parts 46–49
    by substitution 7
    constant
    of exponential functions
    of products
    of trigonometric functions
    of quotients 1–2
    of the product of
    hyperbolic functions 37–38

Integration of the squares
    of the hyperbolic function 36

Integration using t-formulae 103

INDEX

Inverse circular function 49–50

Inverse Hyperbolic Functions 33–34

Integrating factor 80

Length of arc 92
    in cartesian 92
    parametric 92–93
    in polar 92

Logarithmic
functions integration of 21

Mean value 111

Mid-ordinate rule 64

Moment
    first of area 74
    first of volume 75

Osborne's rule 30

Particular integral 85–89
    failure case 86

Reduction formulae 52

Root mean square value 112

Quadratic functions 5

Second order linear differential 84

Separation of variables 77

Sign of the area under the curve 10

Simple Differential Equation 3

Simpson's rule 66

Surface of revolution 98

Substitution, integration by 7

Trapezoidal rule 63

Trigonometric
    function integration of 15

Variable separable 77
    velocity 6

Volume of revolution 72